Water and Western Energy

Studies in Water Policy and Management
Charles W. Howe, General Editor

Water and Western Energy: Impacts, Issues, and Choices

Steven C. Ballard, Michael D. Devine, and Associates,
SCIENCE AND PUBLIC POLICY PROGRAM, UNIVERSITY OF OKLAHOMA

Will the lack of adequate water supplies constrain energy development in the west? The authors of this book stress that, although technically enough water may exist to support high levels of energy development, answers to this question based on a simple comparison of water needs versus supplies may provide a misleading picture. Conflicts over the appropriate uses of limited water resources together with a variety of social and political uncertainties will probably serve as major long-term constraints to western energy development.

In exploring the relationship between western water and energy development, the authors summarize the resource and technological alternatives likely to affect energy development in the region and assess the key social, economic, political, and administrative water-related issues associated with development. They also inform policymakers about options (among them conservation, augmentation, water quality protection, management, and changing intergovernmental responsibilities) for dealing with these issues. They emphasize that future policy and research efforts must address explicitly the close relationship between the quantity and quality of water. Once this is done, they maintain, several technological and institutional alternatives exist for addressing water resource issues.

Steven C. Ballard is assistant professor of political science and assistant director of the Science and Public Policy Program at the University of Oklahoma. Michael D. Devine is director of that program and a professor of industrial engineering at the university. The Science and Public Policy Program is an interdisciplinary research group that has been active in energy and environmental policy for over a decade.

Water and Western Energy: Impacts, Issues, and Choices

Steven C. Ballard, Michael D. Devine,
Michael A. Chartock, Martin R. Cines,
Constance E. Dunn, Carolyn M. Hock,
Gary D. Miller, Larry B. Parker,
David A. Penn, George W. Tauxe

SCIENCE AND PUBLIC POLICY PROGRAM
UNIVERSITY OF OKLAHOMA

Studies in Water Policy and Management, No. 1

Westview Press / Boulder, Colorado

Studies in Water Policy and Management

Copyright © 1982 by Westview Press, Inc.

Published in 1982 in the United States of America by
    Westview Press, Inc.
    5500 Central Avenue
    Boulder, Colorado 80301
    Frederick A. Praeger, Publisher

Library of Congress Cataloging in Publication Data
Main entry under title:
Water and western energy.

    (Studies in water policy and management ; v. 1)
    "Published in cooperation with the Science and Public Policy Program,
University of Oklahoma"--Prelim. p.
    Bibliography: p.
    Includes index.
    1. Water resources development--Government policy--West (U.S.) 2. Energy
policy--West (U.S.) I. Ballard, Steven C. II. Devine, Michael D. III. Uni-
versity of Oklahoma. Science and Public Policy Program. IV. Series.
HD1695.A17W37   333.91'00978   81-21978
ISBN 0-86531-332-6   AACR2

Composition for this book was provided by the authors
Printed and bound in the United States of America

To Don Kash and Jack White
who were instrumental in establishing
the Science and Public Policy Program
at the University of Oklahoma.

Their vision and effort provided the organizational support
and intellectual environment for conducting
this interdisciplinary study.

Contents

ix

Table of Contents (continued)   Page

Table of Contents (continued)

Tables

List of Tables (continued)   Page

List of Tables (continued)   Page

Figures

Foreword

This volume dealing with the water problems likely to be associated with western energy development initiates the new Westview series "Studies in Water Policy and Management." The series will address contemporary water problems in the United States and abroad by expanding and clarifying the range of policy alternatives available for solving these problems. It will emphasize the economic, legal, political, and administrative dimensions of water systems and is intended not only for scholars but also for technical and political decision makers.

This book focuses on water availability and water quality problems associated with western energy development, emphasizing the complexity of the legal/administrative system, uncertainties over federally reserved water, pollution from energy facilities, and salinity control. Five categories of policy options are evaluated: conservation, supply augmentation, water quality protection, improved water management by the states, and alternative roles for federal and state governments. Comparisons with the findings of other studies are then made.

Charles Howe
General Editor

Charles W. Howe, editor of Westview's Water Policy and Management Series, is a professor of economics at the University of Colorado. He has served as director of the Water Resources Program, Resources for the Future, Inc., and on the Board of Editors of Land Economics, the American Geophysical Union Water Monograph Series. Dr. Howe's previous books include Natural Resource Economics: Issues, Analysis, Policy and The Management of Renewable Resources in Developing Countries (Westview, 1981).

Acknowledgments

This study was undertaken as part of "A Technology Assessment of Western Energy Resource Development," (contract number 68-01-1916) for the Office of Environmental Engineering and Technology (OEET), for the Office of Research and Development, U.S. Environmental Protection Agency (EPA) by an interdisciplinary research team from the Science and Public Policy Program (S&PP), University of Oklahoma. This study is one of several conducted under the Integrated Assessment Program established in 1975 by the OEET (formerly the Office of Energy, Minerals, and Industry).

Several publications present the major findings and background material on the complete study. A description of the six energy resources studied and their development alternatives is reported in six volumes of our Energy Resources Development Systems Report (White et al. 1979a). Identification of the range of impacts likely to occur at specific sites and region-wide in the eight-state study area can be found in our two-volume Impact Analysis Report (White et al. 1979b). Identification of problems and issues and analysis of policy alternatives for realizing the benefits of energy development while minimizing the costs and risks is presented in our Policy Analysis Report (White et al. 1979c). Finally, the book, Energy From the West, summarizes the major findings from this work (S&PP 1981).

The Project Director during the preparation of this study was Michael D. Devine, Director of the S&PP Program and Professor of Industrial Engineering. Co-Principal Investigators were Steven C. Ballard, Assistant Director of S&PP and Assistant Professor of Political Science, and Michael A. Chartock, Research Fellow in S&PP and Associate Professor of Zoology. Other S&PP team members who co-authored this study are: Martin R. Cines, Research Fellow (Physical Chemistry); Constance E. Dunn, Graduate Research Assistant (Geography); Carolyn M. Hock, Research Fellow and Assistant Professor of Geography; Gary D. Miller, Graduate Research Assistant, now Assistant Professor, Civil Engineering and Environmental Science; Larry B. Parker, Graduate Research Assistant (Political Science) now with the Congressional Research Service of the Library of Congress; David A. Penn, Graduate Research Assistant (Economics); and George W. Tauxe, Research Fellow and Associate Professor of Civil Engineering and Environmental Science.

Allyn R. Brosz, now Assistant Professor of Political Science at Virginia Polytechnic Institute and State University; and Michael P. O'Hassen, former Graduate Research Assistant (Political Science) were members of the S&PP research team during initial stages of this work and prepared background papers used in early drafts of chapters 7 and 8.

Martha W. Gilliland and Linda B. Fenner of Energy Policy Studies, Inc., Omaha, Nebraska, assisted the study by reviewing earlier drafts of this report and by preparing a subcontract report, "Alternative Water Management Strategies for the San Juan River Basin of New Mexico." David S. Lieberman, Research Fellow in S&PP, and Professor of Chemical Engineering and Materials Science participated as a team member in early stages of this study. The EPA Project Officer was Paul Schwengels, Office of Environmental Engineering and Technology, Office of Research and Development. Gene Reetz, U.S. Environmental Protection Agency, Region VIII, Denver, Colorado also assisted greatly by reviewing copies of early draft reports and papers.

Special recognition goes to Irvin L. (Jack) White, Project Director for the first three years of the technology assessment. Jack White was instrumental in defining water resource issues and their relationship to energy development and in structuring the analysis of alternative policy approaches.

This research project was greatly assisted by a dedicated administrative and research support staff. This staff is headed by Mary Zimbelman, Assistant to the Director. Martha Jordan, Librarian, and Ellen Ladd, Clerical Supervisor, played crucial roles in organizing reference materials, and in editing and producing the final manuscript. Ellen Ladd's tireless efforts to produce a high quality manuscript are greatly appreciated. Lennet Bledsoe was responsible for typing all of the final manuscript. Virginia Newman, Graphic Artist, was responsible for the majority of the graphics. Rod Hedges assisted in producing the manuscript and in preparation of the graphics.

The research team was also greatly assisted by an Advisory Committee with a wide range of expertise and interests in water resource issues. The members of this committee, and their organizational affiliations at the time of their participation, were: Jack A. Barnett, Western States Water Council; Phillip M. Burgess, Western Governors' Policy Office; Lee C. Dutcher, U.S. Geological Survey; George Pring, Environmental Defense Fund; Frank Odasz, Energy Transportation Systems, Inc.; Kenneth Kauffman, U.S. Bureau of Reclamation; Joyce Kelley, U.S. Department of the Interior; W. F. Lorang, El Paso Natural Gas Company; Dean Mann, University of California at Santa Barbara; J. William McDonald,

Colorado Water Conservation Board; John Stencel, Rocky Mountain Farmers Union; Vernon Valantine, Colorado River Board of California; and Robert Siek, Council of Energy Resource Tribes. Although the Advisory Committee has provided invaluable assistance, this does not imply that they concur in the opinions expressed herein.

Although the work upon which this book is based was supported in part by the U.S. Environmental Protection Agency, the opinions, findings, and conclusions expressed do not necessarily reflect the view of the EPA. This book is the sole responsibility of the Science and Public Policy Program of the University of Oklahoma.

Steven C. Ballard
Michael D. Devine

Introduction

This study analyzes the relationship between energy development and water resources in eight western states. It has two purposes: (1) to describe the most important water use and water quality impacts associated with energy development; and (2) to identify and evaluate policy alternatives for addressing the problems and issues. The book is divided into two parts. Part I (Chapters 1-3) presents the purposes and approach of the study and describes the water quality and availability problems caused or aggravated by energy development. Part II (Chapters 4-9) describes and evaluates policy options for dealing with these problems. The final chapter (9) provides a summary of findings and makes a brief comparison with two other recent studies.

Two characteristics of the work are important in understanding its conclusions. First, the analysis is future oriented but not predictive. That is, its purpose is to assess potential problems if certain levels of energy development occur, but no attempt is made to predict energy development patterns over the next four decades. Second, the study takes a broad, interdisciplinary approach to water resource issues, requiring that social, economic, technical, legal, and environmental perspectives be integrated.

IMPACTS AND ISSUES

Western energy development decisions will be entangled in a number of water use and water quality protection issues, many of which will exist even without energy development. Not only will energy development aggravate these problems but they, in turn, can directly influence energy resource development decisions.

Water Requirements For Energy Development

Several uncertainties surround estimates of future water requirements for energy, including a lack of experience with new synthetic fuel technologies, variations in the type of cooling process that could be chosen, and the inherent difficulties in estimating future energy development levels in the region over the next several decades. On a regional level, it is clear that enough water physically exists to support large levels of energy development. For example, the total water requirements for our Nominal Demand scenario, representing a high level of development by the year 2000, would consume no more than 22 percent of the remaining unused water supplies in the Upper Colorado and Upper Missouri river basins. Nevertheless, water is relatively scarce in many energy-rich regions, including the Powder River Basin in Wyoming and Montana, the oil shale area of western Colorado and eastern Utah, and the Four Corners area of New Mexico. In addition, a variety of institutional and political questions will determine the amount of water available for new energy projects as indicated below.

Pollution From Energy Facilities

Even if current discharges are met, energy production and conversion processes can pollute both surface and groundwater either through the disposal of waste products or through the disruption and contamination of aquifers during mining and in situ recovery operations. Given current federal and state regulations and the level of uncertainty, existing environmental regulations are inadequate to ensure that long-term or irreversible damage to surface and groundwater quality does not occur.

Increasing Demands For Water Use

Although agriculture is the dominant water user in the study area, there is a rapidly growing demand for water from other users. These demands threaten many existing interests and values which have developed within a predominantly agricultural economy. In addition, disputes among energy, environmental, Indian, agricultural, municipal, and other interests establish an increasingly complex context for water policy.

Uncertainty and Complexity Of The Water Policy System

The institutional system for managing water resources includes a combination of state water law, federal water policies, court cases, interstate agreements, and international treaties which determine how water will be used. Three of the most serious problems with this system are: (1) state appropriation systems are designed primarily to react to questions about the legality of rights and uses and not to manage the resource on a day-to-day basis; (2) the water allocation system in most western states is not tied directly to water quality protection; and (3) few incentives exist for conserving water, as exemplified by the "use or lose" and "nonimpairment" doctrines.

Reserved Water Rights

When the U.S. establishes a federal reservation such as a national park, military installation, or Indian reservation, a sufficient quantity of water is "reserved" to accomplish the purposes for that particular land. This reserved rights doctrine is significant because the federal government and Indian tribes own large amounts of western land, for example, about 70 percent of the land in the Colorado River Basin. At the present time relatively little water has been put to use under this doctrine; however, since these rights could be exercised for large quantities of water, uncertainty exists as to the amount of water available under state appropriation systems.

Salinity Control

Salinity has been singled out for regulatory control by the federal government and each of the states in the Colorado River Basin and is of increasing concern in the Yellowstone River Basin. The major sources of salinity currently are natural salt flows and runoff from irrigated agriculture. Energy development can magnify salinity problems by increasing salt loading in runoff from mining areas and due to the concentrating effect of consumptive water use. Thus, energy development is likely to intensify salinity control problems.

POLICY ALTERNATIVES

There are a variety of technical, economic, and institutional approaches to deal with the problems and issues identified. This book discusses five categories of alternatives:

(1) conservation,

(2) augmentation,

(3) water quality protection,

(4) state administration and management, and

(5) regional and federal roles.

Within each category, several alternatives are identified and evaluated using a range of criteria--effectiveness, efficiency, equity, flexibility, and feasibility.

Water Conservation

Alternatives for conserving scarce water resources include minimizing use in energy conversion facilities and improving agricultural irrigation practices. Conservation in energy development appears most promising in reducing potential conflicts. Changes in process designs of synthetic fuel plants can substantially reduce consumptive water requirements. Water-saving cooling technologies can also substantially reduce water requirements. Wet/dry cooling of power plants can reduce water requirements by about 70 percent, while intermediate wet cooling of coal synfuel plants reduces them by 20 to 30 percent. Although these water savings can increase economic costs, the economic penalty for intermediate wet cooling is expected to be less than 0.5 percent in synfuel plants and from 3 to 5 percent in power plants.

Considerable uncertainty exists about how much water can be "saved" by alternative irrigation practices because of questions about the amount and nature of irrecoverable losses. Nevertheless, the water "saving" potential is not as significant as the potential for improved water quality. Improvements in irrigation efficiency will face both economic and institutional obstacles. Generally, the economic costs far exceed the benefits to the individual farmer. If this option is to be implemented on a large scale, government subsidies would be required.

Augmentation of Supply

Water supplies can be augmented by a variety of means, including the use of saline surface water or groundwater by energy conversion facilities, increased use of groundwater, weather modification, removal of nonproductive vegetation along stream beds, and inter- or intra-basin transfers. Use of saline ground and surface waters for energy appears particularly attractive. However, the overall effectiveness of this option is uncertain and requires site-specific studies. Where applicable, it can provide new water supplies to energy conversion facilities at a maximum increased product cost of 2 to 6 percent. Although more waste disposal would be required, this option would have environmental and economic benefits if used as a part of a salinity control program.

Most of the other augmentation alternatives appear to be costly choices for dealing with the complex problems and issues affecting western water resources. Substantial quantities of water can be added by these alternatives; however, if economic costs, environmental damage, and long-term ecological risks are considered, they are likely to worsen rather than improve conflicts over the appropriate use of water.

Water Quality Protection

Options to protect water quality include improved water quality controls for energy development, temporary sewage treatment measures for energy "boomtowns", various salinity control projects, and a salinity offset policy for energy projects. In addition, several of the water supply and conservation alternatives have implications for water quality, especially the use of saline water in energy facilities and improved irrigation practices.

Improved water quality controls for energy development activities consist of three primary elements: predevelopment monitoring of water quality, water quality control plans, and monitoring and research. These options would be an expanded and more comprehensive approach to already ongoing activities. Their goal would be to ensure that best available practices are used in the short-term and that over the longer term the knowledge about geohydrology and the behavior of potential contaminants is improved. This will ultimately improve our ability to define the risks of energy development and the effectiveness of various control measures.

State Administration and Management

Four policy alternatives are discussed for increasing the states' capacity for managing water resources: reductions in institutional constraints to water transfers, water rights exchange systems, time-limited permit systems, and coordinated administrative structures.

These alternatives would help to meet expanding demands for water, and help to improve day-to-day management of water resources. An important difference among the options is whether responsibility for meeting multiple uses will rest primarily with state governments or with the private sector. Public-operated water banks would provide a large degree of state control. In contrast, reducing institutional barriers to water transfers would gradually allow the market system to play a greater role in how water is used.

A central weakness of current state administrative systems is that water availability and quality are the responsibility of different state agencies. Coordinated administrative systems are specifically designed to address this problem. This option could take a variety of forms, including establishment of Departments of Water Resources which would be mandated to balance availability and quality concerns. Whatever the form, a clear need exists in every state of the study area to address this problem.

Feasibility is a concern with almost any change in existing institutional systems. However, many of these choices have already been introduced in states facing water problems. For example, Colorado, Montana, New Mexico, and Arizona have changed policies regarding beneficial use to account for instream values. Because of increasing demands on water and pressures to improve management of the resource, such approaches to water administration are likely to become more feasible in the near future.

Regional and Federal Roles

Policy alternatives considered in this category are: increased federal technical assistance to states, quantification of federal reserved rights, "real-cost" water pricing, and federal-interstate compacts. Among the most important factors affecting the need for such approaches are: (1) the vast energy resources located in water-scarce areas; (2) the high percentage of federal lands in the study area; and (3) the complexity of the water policy system.

Of these alternatives, federal technical assistance in developing water information systems and federal-interstate compacts have several advantages. Among the most important of these is their potential to improve the information about water availability, existing uses, and relationships between quantity and quality. Improving the information base in these areas is a prerequisite for several other management goals. The advantage of federal technical assistance is that it can increase states' day-to-day management capacity with a

minimum of federal "interference". The federal-
interstate compact provides a more integrated approach
to developing water resource plans, in addition to
giving the states a strong voice in the compact commission.

POLICY DIRECTIONS

The western water situation has changed rapidly as
demands for a scarce resource continue to increase from
energy development, national defense uses, mineral and
industrial expansion, reserved rights, instream values,
salinity control, and population growth. The implication of this growth is that the water policy system,
which has dealt successfully with many water problems in
the past, must continue to adapt to changing circumstances in order to balance the competing needs for
water in the future. Two general conclusions result
from our study. First, water resource management should
continue to be primarily a state responsibility; there
is little evidence suggesting that a centralized or
uniform federal approach to water policymaking is appropriate. Second, if states are to be successful in
resolving water resource conflicts, three policy guidelines should be followed: (1) policies should be flexible so that the range of legitimate, multiple uses of
water is recognized; (2) policies should encourage water resource management to enhance day-to-day control
and integrate water quantity and quality concerns; and
(3) policies should enchance public education and participation in water resource policymaking.

Part I

Energy Development and Water:
Impacts and Issues

1
Water and Energy Development in the West

INTRODUCTION

If the U.S. is to decrease its dependence on imported oil, then it must use its available energy supplies more efficiently and increase domestic energy production. Given the substantial energy resources in the western U.S., this region is expected to be a major contributor to increased domestic supplies. The production of uranium and coal has increased greatly in the past five years; new discoveries of oil and gas in the overthrust belt will almost certainly result in increased development in Montana, Idaho, Wyoming, and Utah; and a major portion of synthetic fuel development is likely to occur in the oil shale and coal areas of the West.

Water availability and quality will be among the most critical problems associated with expanded western energy development. Although water has always been scarce in the West, enough has generally existed to provide supplies to a substantial number of users, primarily irrigated agriculture and municipalities. However, the central question regarding future development of the region is whether or not enough water exists to support traditional users and the growing demands of energy development, other industrial development, defense installations, Indians, environmental interests, and others.

The importance of water to western life has always been well-recognized. A critical consideration in this region,[1] in contrast to most of the remainder of the country, is that the supply is relatively small and highly variable. As shown in Figure 1-1, precipitation levels range from less than 10 inches per year in the desert southwest to 10 to 20 inches in the Northern Great Plains. The two major river basins in the eight-state area, the Colorado and Upper Missouri, provide basic water supplies for ten states, extending from the Dakotas through the Rocky Mountains to the desert southwest, including southern California. Although considerable groundwater exists in this region, much uncertainty exists about its quantity, quality, and location in relation to where it might be needed.

3

4

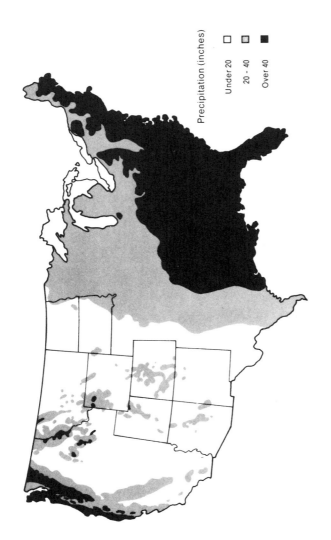

Precipitation (inches)

☐ Under 20

▨ 20 - 40

■ Over 40

Figure 1-1:   Annual Precipitation

Three aspects regarding these resources are critical to western energy development. First, a large percentage of surface water is already committed to use. This is particularly true in the Colorado River Basin (CRB) --in fact, some states have granted permits to more water than users may ultimately be able to use under existing laws. Thus, new water-right applications, even if they represent only a very small part of a state's total supply, will be competing for a very small quantity of uncommitted supplies.

A second aspect is that energy resources are often located disadvantageously with respect to water supplies. Perhaps the best example of this is the oil shale region of western Colorado where approximately 85 percent of the nation's high-grade recoverable oil shale reserves are located in a two-county area. This concentration of a resource occurs in an area where most streams are fully used and where flowing water is particularly valued for environmental purposes, such as protecting instream water uses. Locational problems also exist in the Upper Missouri Basin, although considerably more water exists in the Upper Missouri than in the Colorado. The richest coal deposit in this area is the Powder River Basin in southeastern Montana and northeastern Wyoming. Limited water for energy development appears to exist in the Tongue, Powder, Bighorn, or Belle Fourche rivers, which are nearest to these coal resources (see Figure 2-1 in Chapter 2). Thus, water for energy development will probably have to come from the Yellowstone or Missouri mainstem, requiring transfer and storage projects.

The third, and perhaps most important, factor is that a new group of interests is aggressively pursuing and acquiring rights to use water in ways or in quantities that formerly had not been recognized, or received low priority. Water rights to develop energy resources, meet the needs of Indians, protect water quality, preserve instream values such as recreation and aquatic life, and preserve Wild and Scenic Rivers have been recognized and, in many cases, threaten to reduce supplies for irrigated agriculture and municipalities. Based on this, it is easy to understand why water availability and quality questions generate intense conflicts in the West--resolution of the issues will have a profound effect on the basic economic, cultural, and social character of the region.

As a result of this situation, western energy development decisions will be entangled in a number of water use and water quality protection issues:

- The nature and complexity of the water policy system creates barriers to change, discourages a diversity of water uses, and provides few incentives for conservation of the resource.

- The growing demand for water among a diversity of users is leading to water shortages in some areas and increasing conflicts over the most appropriate uses of this relatively scarce resource.

- Federal and Indian reserved water rights create uncertainty over the amount of water available to other users under state appropriation systems.

- Disputes exist among various interests and institutions over western water management and development, including the extent to which western water and other resources should be used to meet national energy needs.

- Energy production and conversion could pollute surface and groundwaters; given the current level of uncertainty, existing regulations are inadequate to ensure that long-term or irreversible damage does not occur.

- Although energy activites are expected to cause only small increases in salinity levels relative to other sources, energy production levels could be affected by salinity control policies, especially in the CRB, if adequate controls are not established.

- Rapid population fluctuations in small western communities affected by energy development can create water pollution due to inadequate treatment of municipal sewage.

In summary, water availability and quality problems in the West are already serious and are likely to become worse even without energy resource development. Although energy development is not the major cause of most water problems, it is likely to exacerbate many of them. Thus, energy development could be constrained in many areas of the West because of these conflicts. This situation highlights the need for new policies to address these problems and issues.

PURPOSE OF THIS STUDY

This study is an extension of our larger study, Energy From the West (White et al. 1979a, 1979b, 1979c and S&PP, U. of Okla. 1981) which analyzes a broad range of impacts and issues likely to result from western energy development and identifies and evaluates selected policy alternatives for dealing with the most serious issues. In this previous work, we examined problems and issues in the following categories:

- Water availability,

- Water quality,

- Air quality,

- Land use and reclamation,

- Housing,

- Growth management,

- Capital availability,

- Energy transportation, and

- Energy facility siting.

This study relies heavily on Energy From the West for identifying the major water-related impacts and issues. The purpose of this report is to: (1) update the information on impacts and issues (chapters 2 and 3), and (2) extend the analysis of policy alternatives for dealing with energy-related water availability and water quality issues (chapters 4-8). The remainder of this chapter provides an overview of the energy resources and development technologies considered, and the approach to the study.

ENERGY RESOURCES AND DEVELOPMENT ALTERNATIVES

The impacts and issues associated with developing six energy resources--coal, crude oil, natural gas, oil shale, uranium, and geothermal--are considered. The quantities of reserves for these resources are shown in Table 1-1. Approximately 36 percent of all U.S. coal is located in the study area; virtually all the nation's high grade oil shale is located in the Green River Formation in western Colorado, Utah, and Wyoming; and almost

TABLE 1-1: Proven Reserves of Six Energy Resources
in the Eight-State Study Area

Resource	Western Reserves (10^{15} Btu's)	Percent of U.S. Total Reserves
Coal	3,430.0	36
Crude Oil	12.2	6
Natural Gas	19.9	8
Oil Shale (high grade)	464.0	100
Uranium	246.6	90
Geothermal[a]	650.0	22

Source: White et al. 1979c

[a]This figure includes reserves, submarginal resources, and paramarginal resources.

all of the nation's high grade uranium ore is located in the eight states, primarily in New Mexico and Wyoming.

When the western energy resources described above are developed, the effects on the West and the rest of the nation will vary depending upon the level, rate, pattern, and technological alternatives chosen. This study considers a range of development technologies for each resource, as indicated in Table 1-2. The technologies selected are representative of those likely to be used to develop western energy resources during the next twenty-five years. A brief summary of these technologies follows. (More detailed descriptions can be found in White et al. 1979a.)

Coal Development Technologies

Technologies included for producing and converting coal to other energy forms are coal mining, coal-fired steam-electric power plants, Lurgi and Synthane coal gasification, Synthoil coal liquefaction, and unit-train and slurry-pipeline transportation.

Coal Mining. Both surface and underground coal mining are considered in this study. Surface (or strip) mining begins with the removal and storage of topsoil from the area. (Topsoil is stored so that it can be replaced later during reclamation.) The overburden--that is the rock and soil material between the surface

TABLE 1-2:   Development Alternatives

Coal:	Surface and Underground Mining Direct Export by Unit Train and Slurry Pipeline Electric Power Generation Gasification Liquefaction Transportation by Pipeline High-Voltage Transmission
Oil Shale:	Underground Mining Surface Retorting Modifed In Situ Transportation by Pipeline
Uranium:	Surface, Underground, and Solutional Mining Milling Transportation by Truck
Oil and Natural Gas:	Conventional Drilling and Production Enchanced Oil Recovery Transportation by Pipeline
Geothermal Energy:	Hot Water Production Electric Power Generation High-Voltage Transmission

---

and the coal seam--is loosened by blasting and removed using a dragline. The dragline lifts the overburden and places it on a spoils pile adjacent to the mining area. The exposed coal is then mined and loaded into large trucks or onto conveyor belts for transportation to either a conversion or a loading facility. The present practice is for mining and reclamation to proceed simultaneously. The overburden is placed in the mined out area, graded, and contoured; topsoil is then replaced and the area is revegetated.

Room-and-pillar mining is the predominant underground coal mining technique. As the coal is mined, pillars of it are left in place to support the mine's roof (roof supports are used in addition to the pillars). Mechanical continuous miners are used to scrape the coal from the seam and load it directly onto a conveyor or into a rail car. Reclamation for underground mines involves permanently disposing of the spoils mined along with the coal and of the material removed to gain access to the seam. These waste materials are usually stabilized with lime and deposited in sealed landfills.

Coal-Fired Steam-Electric Power Plants. The coal-fired power plants analyzed are the common steam-electric type. An overall efficiency for the plant of 34 percent is assumed, including environmental controls. These environmental controls include a wet limestone scrubber for flue-gas desulfurization (FGD) and an electrostatic precipitator for particulate removal. Because of thermodynamic limitations, almost two-thirds of the heat generated in a power plant boiler must be dissipated. For the basic plant, it is assumed wet cooling towers would be used to dissipate the heat by evaporating water into the atmosphere. To examine the sensitivity of impacts to the cooling technology, a wet-dry cooling system is also examined.

High-Btu Coal Gasification. Two high-Btu coal gasification processes are considered. The Lurgi process was selected for study because it is a presently available commercial-scale technology. The Lurgi gasifier is in commercial operation today in South Africa. The Synthane process was selected as representative of a number of second-generation processes which could be commercially available by 1985 to 1990.

In high-Btu coal gasification, coal is transformed into gas by heating it in the presence of oxygen and steam to produce carbon monoxide and hydrogen. This mixture of gases is then upgraded to create synthetic natural gas (primarily methane) in a separate reactor using a catalyst. Water is used in these facilities both for the gasification process and for cooling. As with the steam-electric power plant, it is assumed wet cooling towers would be used in the basic plant, but wet-dry and all-dry cooling are also examined.

Coal Liquefaction. Coal liquefaction processes are at an earlier stage of development than gasification, and therefore data on liquefaction processes are somewhat limited and uncertain. The liquefaction process considered in this study is the Synthoil process developed by the Bureau of Mines.[2] Water is used both for cooling and as a source of hydrogen for the process.

Coal Transportation. Two options for transporting coal are analyzed: unit trains and coal-slurry pipelines. For the purposes of this water study, only the slurry pipeline is of interest since train transportation involves negligible amounts of water. In a coal-slurry pipeline, the coal is pulverized, mixed with water, and pumped through a pipeline. Approximately equal parts (by weight) of coal and water are required. It is assumed that the carrying capacity of a "typical" coal-slurry pipeline is 25 million tons per year (MMtpy) of coal.

Oil Shale Development Technologies

The technologies considered for developing oil shale are underground oil shale mining and surface retorting using the TOSCO II process, and modified in situ recovery using the Occidental process.

Underground Oil Shale Mining. Conceptually, underground oil shale mining is similar to underground coal mining and most frequently uses the room-and-pillar method. Compared to coal mines, however, oil shale mines are very large, with roof heights of as great as 60 to 80 feet. These large rooms are mined in two zones. The top zone is mined with equipment extracting the shale from the wall (or face) of the resource, while the bottom zone is mined by extracting the shale from the floor (or bench). Large front-end loaders are used to load the mined shale into trucks which transport it to a sizing and crushing facility. Because of the enormous size of these mines, equipment more commonly seen in surface mines, such as large trucks and drill rigs, is used. A schematic drawing of an underground oil shale mine is shown in Figure 1-2.

Surface Oil Shale Retorting. TOSCO II is the surface retorting process assumed. In the TOSCO II retort, one-half inch diameter ceramic balls are heated to about 900°F and then put into the retort with small pieces of raw shale. The retort vessel is then rotated so that the balls heat the shale by contact and, at the same time, crush it to a powder. The oil is collected, and the pulverized spent shale is separated from the balls with moving screens and carried away for disposal. A low-Btu gas also is generated; it is collected and used as a fuel in the heater for the ceramic balls. Because the energy content of shale is relatively low (only about 25 to 35 gallons of shale oil can be extracted from a ton of ore), a large quantity of spent

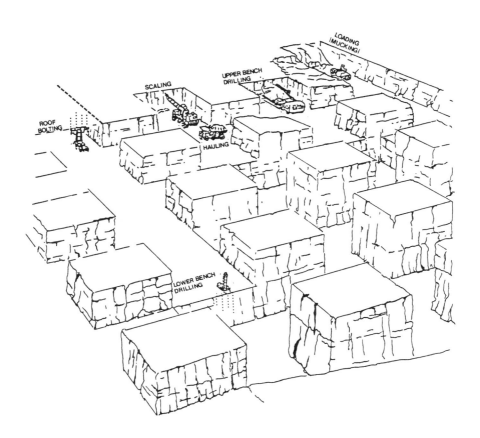

Figure 1-2:   Oil Shale Room-and-Pillar Mining

shale must be disposed of. The pulverized spent shale from the TOSCO II process can be set as a cement with about 13 percent water. It is assumed that a small canyon in the vicinity of the retort would be used for shale disposal and filled with the shale cement to a depth of several hundred feet. The disposal area would then be covered with topsoil and vegetated.

In Situ Oil Shale Retorting. Also analyzed is the Occidental modified in situ retorting process (see Figure 1-3). In the modified in situ process, oil shale is mined from a region below a high-quality ore deposit. The mined oil shale can either be discarded or processed in a surface oil shale retort. The highgrade ore is then broken into rubble with shaped-charge blasting to form large in situ retorts, measuring more than 100 feet on a side and nearly 300 feet high. Air and steam are circulated through the shale rubble, and the shale is heated until it ignites in a burning front which moves downward, releasing shale oil ahead of the combustion. The shale oil is collected in sumps, and the gases produced are treated to recover additional energy and to reduce emissions. In situ retorting requires the handling of a much smaller quantity of shale than surface retorting. As for disposal, the raw shale from in situ retorting is more stable than processed shale from surface retorting.

Uranium Development Technologies

The technologies for producing uranium include surface, underground, and solutional uranium mining; and acid-leach milling of the ore from surface and underground mines to produce "yellowcake" (uranium oxide $[U_3O_8]$).

Surface and Underground Uranium Mining. A surface uranium mine is more similar to pit mining than to the strip mining typically used to recover coal. Uranium is located in veins which vary in ore quality and are irregular in location. Mining is therefore carried out selectively, with lower quality ore left as a spoil. As truckloads of ore leave the pit, radioactivity is measured to grade the ore for milling. The equipment used for mining uranium is similar to that used for mining coal but is usually smaller in size because of the smaller quantities of ore involved.

Reclamation of uranium mines includes layering the overburden by composition to limit the potential for damage from trace elements in the overburden leaching

14

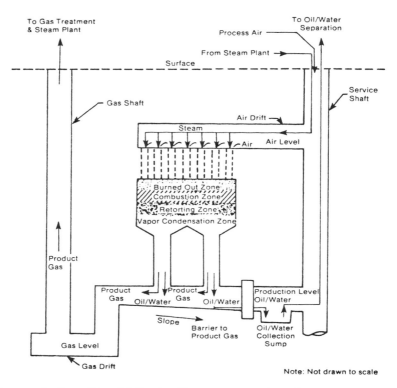

To Gas Treatment
& Steam Plant

To Oil/Water
Separation

Process Air

From Steam Plant

Surface

Service
Shaft

Gas Shaft

Air Drift

Steam

Air Level

Air

Burned Out Zone
Combustion Zone
Retorting Zone
Vapor Condensation Zone

Product
Gas

Product
Gas

Product
Gas

Production Level
Oil/Water

Oil/Water

Oil/Water

Slope

Gas Level

Barrier to
Product Gas

Oil/Water
Collection
Sump

Gas Drift

Note: Not drawn to scale

Figure 1-3: Simplified Sketch of In Situ Oil Shale
Retort

Source: Adapted from Ashland and Occidental 1977.

into groundwater and to assure the stabilization of
materials that may be radioactive.

Underground uranium mining is done by a room-and-
pillar method.   Ore is mined using continuous miners and
removed by conveyors to the access shaft where self-
dumping buckets are used to haul the ore out of the
mine.

Solutional Uranium Mining. Solutional uranium mining
is included in this study to present a comparison with
surface mining and also because the irregular seams, or
veins, and the nature of the geology of uranium make this
the most economical mining technique.   In solutional mining an alkaline solution is injected into the ore formation through a group of injection wells.   This solution

dissolves the uranium compounds; the "pregnant" solution of uranium is pumped out of the formation in recovery wells; and the solution is then processed to recover yellowcake. Thus, the mining and milling processes are combined in solutional mining.

Reclamation after solutional mining involves cleansing the groundwater in the mined-out area to ensure that uranium and other heavy-metal compounds do not remain. This is accomplished by pumping clean water into the formation and recovering the remaining alkaline solution and dissolved compounds. The recovered solution is then either cleaned or disposed of in an evaporative pond.

Uranium Milling. Uranium milling is a process by which uranium is extracted from ore and concentrated into a yellowcake which is about 90 percent U_3O_8. The milling process included in this analysis is acid leaching. The ore is crushed, slurried, and pumped into heated, agitated tanks where sulfuric acid is added. The uranium ore dissolves in the acid, and the solution is then separated from the waste materials, or tailings, which are discarded in a tailings pond. The solution is concentrated with additional chemical processes, and finally the U_3O_8 is separated out of the solution and dried for shipment.

Oil Development Technologies

Conventional Oil Production. In this study, oil production is assumed to include rotary drilled wells in which water-based mud is used as a drilling fluid for lubrication and to remove cuttings. The wells are lined with steel-pipe casing cemented in place and the oil is produced through tubing going from the bottom of the well in the oil reservoir to the surface. No wellhead processing of the oil is considered except for water removal, after which the oil is piped out of the region for refining elsewhere.

Enchanced Oil Recovery. The enchanced oil-recovery method analyzed is steam flooding. Two sets of wells are used, one to inject the steam and the second to recover the oil-and-water mixture. The steam increases the temperature of the oil so that its viscosity is reduced; thus, it flows more readily. The oil is then pushed by the steam pressure to the recovery wells. Most of the steam injected is recovered as water in a mixture with the oil. This water must be disposed of in evaporation ponds or cleaned and recycled to the steam generators.

Natural Gas Development Technologies

Conventional techniques for the production of natural gas are included. These include rotary drilling using water-based mud as a drilling fluid; wells cased with a heavy pipe cemented in place; and gas production through tubing (of a smaller diameter) from the production zones. The only processing assumed is drying using a glycol process.

Geothermal Development Technologies

Hot-water geothermal development is analyzed because it is one of the most likely geothermal resources to be developed in the eight-state study area. This resource is defined as having a temperature above 300°F; it is also assumed that power is produced by either a flashed-steam or binary-cycle process.

In the flashed-steam process, the hot, high-pressure water is expanded to a low pressure so that it vaporizes. The vapor is then used to drive a low-pressure turbine, which in turn drives an electric generator. If the hot water is impure and contains salts or solid particulates, the flashed steam must be cleaned before entering the turbine.

In the binary-cycle process, the hot water is used to heat another fluid with a lower boiling point (such as propane) which is then vaporized and used to power a turbine. The binary cycle has fewer problems with impure water but is more complicated than the flashed-steam process. In both processes, cooling is assumed to be by wet cooling towers that use the geothermal condensate as a source of water.

STUDY APPROACH

Identification of Impacts

In order to project the impacts of greatly increased energy development on water quality and availability, both site-specific scenarios and regional scenarios were constructed. The location of the site-specific scenarios is shown in Figure 1-4. The six site-specific scenarios combine different local conditions (e.g., topography, meteorology, water availability, current population, and community services and facilities) with alternative energy development technologies. Table 1-3 lists the facilities for each of the six sites. For example, a Farmington, New Mexico, scenario includes surface coal mines, a coal-fired steam-electric power

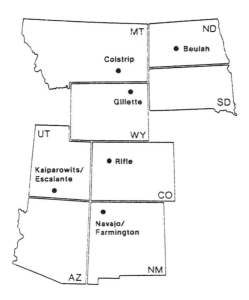

Figure 1-4:   Eight-State Study Area and Six Scenario
              Sites

plant, a Lurgi high-Btu coal gasification facility, a
Synthane high-Btu coal gasification facility, a coal
liquefaction plant, a uranium mining and milling facility, and the necessary energy transportation systems
(high-voltage transmission lines and pipelines).

Two regional scenarios--one based on "Low" and one
on "Nominal" national demands for energy between 1975
and 2000--were created primarily using Stanford Research
Institute's (SRI) interfuel competition model (Cazalet
et al. 1976). However, adjustments to the SRI figures
were made to include in situ oil shale and geothermal
energy development (White et al. 1979b, pp. 914-928).
Table 1-4 summarizes the quantities of energy produced
from each resource for these two scenarios. The "Low"
and "Nominal" demand scenarios correspond to national
energy demands by the year 2000 of 124 Q^3 and 155 Q,
respectively.

TABLE 1-3: Standard Size Facilities in The Six Site-Specific Scenarios

Facility	Navajo/ Farmington New Mexico	Kaiparowits/ Escalante Utah	Rifle, Colorado	Gillette, Wyoming	Colstrip, Montana	Beulah, North Dakota
Coal						
Mine/Export (25 MMtpy)						
Rail				1		
Slurry				1		
Power Plant (3,000 MWe)	1	2	1a	1	1	1
Gasification (250 MMcfd)						
Lurgi	1			1	1	2
Synthane	1			1	1	2
Synthoil (100,000 bbl/day)	1			1	1	
Oil Shale						
Surface Retorting (50,000 bbl/day)			1			
In Situ (57,000 bbl/day)			1			
Uranium						
Mine/Milling (1,000 mtpy)	1			1		
Solutional Mining (250 mtpy)				1		
Natural Gas Wells (3 MMcfd)				83		
Crude Oil Wells (125 bbl/day)			400			

MMcfd - million cubic feet per day   bbl/day = barrels per day   mtpy = metric tons per year

aThe Rifle power plant is only 1,000 MWe.

TABLE 1-4: Low and Nominal Demand Case Scenarios of Western Energy Resource Production

Energy Type	1975	Nominal Demand Case 1980	1990	2000	Low Demand Case 1980	1990	2000
Coal Production[b]	1.57	5.04	12.60	29.90	4.28	9.80	22.10
Direct Use	1.50	3.73	8.30	14.80	3.13	6.38	11.10
Unit Train	1.50	3.14	5.33	7.07	2.71	4.13	5.28
Slurry Pipeline	0	0.52	2.97	7.75	0.42	2.26	5.79
Gasification	0	0	0.23	3.83	0	0.15	2.30
Liquefaction	0	0	0.001	0.31	0	0.001	0.38
Electrical Generation	0.03	0.48	1.08	1.43	0.41	0.84	1.08
Oil Shale	0	0.001	0.95	4.76	0.001	0.19	0.95
Uranium Fuel	1.95	4.77	12.70	23.80	4.15	9.46	17.10
Gas (Methane)	1.71	1.97	2.08	1.06	1.99	1.89	1.19
Domestic Crude Oil	1.43	1.67	1.32	1.03	1.74	1.34	1.20
Geothermal	0	0	0.09	0.15	0	0.01	0.02
Total	6.66	13.45	29.74	60.70	12.16	22.69	42.56

Q's from Eight State Region[a]

[a]Input values for coal direct use; output values for others.

[b]Coal subcategories do not add to total because of energy losses in the energy conversion subcategories.

In the SRI model, the geographical distribution of development was carried out by dividing the western states into only two subregions, the Powder River and the Rocky Mountain areas. For some of our impact analyses, it was necessary to disaggregate further, to an individual state. The number of facilities by state that we used in our analysis is given in Table 1-5. For the near future (up to 1985), disaggregation was done on the basis of the locations of energy facilities that have already been announced (Denver Federal Executive Board and Mountain Plains Federal Regional Council 1975). Thereafter (for 1990 and 2000), development was assumed to be proportional to the proven reserves in each state. Disaggregation based on resource levels was done only to provide a basis for impact analysis. Of course, the actual siting of facilities depends on a number of other factors, including water availability, mining costs, and several legal and insititutional factors.

Policy Analysis

Based on an analysis of the impacts likely to result from developments at the levels specified in the scenarios, critical water issues are identified. For example, energy development will add to existing problems, such as water shortages and increased salinity; and create new problems, such as pollution from waste disposal and the disruption of aquifers. This study attempts to identify the most critical problems and issues--those which affect a wide range of interests and values, which can create long-term or irreversible impacts, or which are likely to be an important influence on the development of the region or on national energy policy.

After describing the issues, a range of policy alternatives to resolve these issues are identified and analyzed. For example, potential problems of water scarcity might be addressed by alternatives which augment supplies (ranging from interbasin water transfers to weather modification), and by alternatives which improve the efficiency of existing uses (such as reducing water requirements for energy conversion facilities). A variety of technological, legal, and institutional options are addressed, which range from incremental changes in state appropriation law to improvements in regional water resource management. Although a wide range of policy options is discussed, the analysis is by no means comprehensive. In addition, inclusion of a policy option should not imply that it is necessarily a good choice for dealing with the problems and issues identified. The overall purpose of the policy analysis is to provide policymakers with information on the costs and benefits of possible courses of action.

TABLE 1-5:   Number of Standard Size Facilities[a]
            by State in the Low and Nominal
            Demand Scenarios

State	1975	1980		1990		2000	
		Low	Nominal	Low	Nominal	Low	Nominal
Colorado							
Power Plants		1	1	1	2	1	2
Modified							
In Situ		0	0	1	3	3	13
Uranium		0	0	1	2	2	3
Natural Gas		0	0	1	8	4	4
Oil Shale		0	0	0	2	2	10
New Mexico							
Natural Gas	19	22	22	21	15	9	8
Crude Oil	7	8	8	6	6	5	5
Uranium	4	9	10	19	22	31	42
Power Plants		0	1	1	1	1	1
Geothermal		0	0	1	10	10	71
Gasification		0	0	0	0	1	2
Utah							
Power Plants	1	1	1	1	2	1	2
Uranium		0	0	0	2	2	3
Oil Shale		0	0	0	0	0	2
Montana							
Power Plants		1	2	3	5	5	6
Gasification		0	0	0	1	9	15
Liquefaction		0	0	0	0	1	1
Wyoming							
Power Plants	3	1	1	3	2	3	3
Uranium		5	6	12	16	22	31
Gasification		0	0	0	0	5	9
Liquefaction		0	0	0	0	1	1
North Dakota							
Power Plants		2	2	4	6	6	9
Gasification		0	0	2	2	13	21

[a]The size of facility is indicated in Table 1-3.

Generally, the emphasis is on analysis of policy alternatives which primarily involve or directly affect energy development; thus, alternatives which receive relatively more attention in this study are water conservation in energy conversion facilities, use of saline water by energy developers, and water quality controls for energy development. Shorter discussions of other augmentation and conservation alternatives are also included so that general comparisons can be made about the relative advantages and disadvantages of several different approaches. In addition, several more comprehensive alternatives are discussed, such as quantification of reserved federal and Indian water rights and regional water institutions; these discussions are intended to take a broader look at water resource management.

In each chapter, alternatives are analyzed in the following steps: (1) a description of the alternative, including its current status; (2) an evaluation, identifying advantages, disadvantages, and feasibility; and (3) a comparison among the alternatives. In order to inform policymakers about the range of trade-offs which can be anticipated if various choices are made, five basic evaluative criteria were used to systematically structure the analysis (See Table 1-6).[4] Each criterion must be defined in terms of the problem which is being addressed. For example, as applied to potential water shortages, effectiveness may be defined in terms of how much water would be saved or added, and whether it would offer a long- or short-term solution. As applied to potential water pollution problems, effectiveness may be defined in terms of how much salinity or toxic substances are removed from a water source.

The basic point about criteria is that while many policymakers desire to have a "bottom-line," no single measure or evaluation criterion can provide an adequate summary of the costs, risks, and benefits of alternative policies. The combination of measures and criteria to be used is determined both by what is being evaluated and the interests and values at stake. Although economic measures and criteria are used most frequently, they are not always applicable and do not always provide an adequate basis for evaluation. For example, dollars are not an adequate measure of aesthetic values nor do they always provide the best indication of how equitably an alternative may distribute costs, risks, and benefits. And while it is possible to determine the dollar cost of environmental controls, the associated social costs often cannot be determined. By themselves, economic measures and criteria can be used to evaluate only one component of overall costs, risks, and benefits.

The evaluation criteria identified in Table 1-6 have been used to help ensure a broad and systematic analysis of policy alternatives. However, not all of these

TABLE 1-6:   Evaluation Criteria

Criterion	What Does It Evaluate?
Effectiveness	Achievement of Objective Does the alternative avoid or mitigate the problem or issue? Is it a short- or long-term resolution or solution? Does it depend on state-of-society assumptions?
Efficiency	Costs, Risks, and Benefits What are the economic costs, risks, and benefits? What are the social costs, risks, and benefits? What are the environmental costs, risks, and benefits? Are they reversible/irreversible, short- or long-term?

Equity	Distribution of Costs, Risks, and Benefits
	Who will benefit? experience costs? assume risks?
	    -Geographically?
	    -Sectorially?

Flexibility	Applicability Are local and regional differences accommodated? Are social and sectorial differences taken into account? How difficult will it be to administer? How difficult will it be to change?
Feasibility	Acceptability/Enforceability Can it be implemented within existing laws, regulations, and programs? Can it be implemented by a single agency or level of government? Is it compatible with existing societal values? Is it likely to generate significant opposition?

criteria are necessarily applicable to each alternative and their relative importance varies from alternative to alternative. Thus, the evaluation in chapters 4-8 is presented in terms of the most important advantages of each alternative, rather than criterion-by-criterion.

ORGANIZATION OF THE STUDY

Chapters 2 and 3 address the water availability and water quality impacts and issues associated with western energy development. Each of these chapters discusses the impact of energy facilities on water and the current or likely conflicts among various interests. These two chapters are followed by a discussion of policy options for dealing with the impacts and issues. In Chapter 4, "Conservation of Water," water conservation in energy development is emphasized, but several choices for agricultural conservation are also included. Chapter 5, "Augmentation of Water Supply," includes a discussion of five alternatives: use of saline water in energy development; water storage and transfer projects; groundwater storage and use; weather modification; and vegetation management. Chapter 6 is devoted to four policy options for protecting water quality: water quality control plans for energy development; temporary sewage treatment measures; salinity control projects; and a salinity offset policy.

The next two chapters present more comprehensive approaches for dealing with western water problems. Chapter 7 looks at water management changes at the state and local level. These include: reducing institutional constraints to water transfers; developing water rights exchange systems; using time-limited permits for new water rights; and improving the administration of water resources. Chapter 8 examines regional and federal water management alternatives, including: expanding federal technical assistance to states; quantifying reserved rights; modifying federal water pricing policies; and establishing federal-interstate compacts. The final chapter (Chapter 9) provides a summary of the report's findings and briefly compares the conclusions of this report with those from two other recent studies.

NOTES

[1]In this study, we look at eight western states: Arizona, Colorado, Montana, New Mexico, North Dakota, South Dakota, Utah, and Wyoming.

[2]Since the time when these technologies were selected for study (late 1975), there has been considerable change in the technical progress on various processes. It now appears that Synthoil will not become commericalized; however, its resource requirements and emissions are similar to other liquefaction methods based on "direct processes" or (hydrogenation) so that the conclusions drawn still apply.

[3]A "Q" or quad is equal to 10^{15} British thermal units. One Q equals approximately 172 million barrels of oil, 60 million tons of western coal, or one trillion cubic feet of natural gas.

[4]For a more detailed discussion of our conceptual approach to policy analysis as a part of technology assessment studies. see White et al. 1979c, Chapter 3.

2
Water Availability:
Impacts and Issues

INTRODUCTION

Although the amounts of water likely to be required by future energy facilities are a small percentage of existing regional availability, large-scale western energy development will raise serious issues concerning the supply and use of water, particularly in the oil shale regions of the Upper Colorado River Basin (CRB), the Four Corners area, and the Powder River Basin in the Northern Great Plains. Competition for water is already keen, and further energy development will add a major competitor, aggravating existing political conflicts over how, and to whom, water resources should be allocated. These conflicts result in part from uncertainties associated with the complex system of appropriated rights, interstate compacts, international treaties, and court decisions, as well as from numerous unanswered questions about prior rights, beneficial uses, and relationships between groundwater and surface water. But the underlying causes of water issues are its scarcity throughout the region, uncertainty resulting from inadequate water-use and resources data, and the multiplicity of political jurisdictions and private sector interests involved.

This chapter describes the water-related issues that are likely to affect expanded energy development in the West. The following section outlines data on physically available water supplies in the eight states. This is followed by an analysis of water requirements for individual energy facilities and for overall energy development in the two major river basins in the region—the Upper CRB and the Upper Missouri River Basin (MRB).

The remaining sections of the chapter discuss four issues likely to be important to future energy development: complexity of the policy system which controls water supplies and uses, increased competition among users, reserved water rights, and jurisdictional disputes.

27

WATER RESOURCES IN THE EIGHT STATES

Figure 2-1 shows the major rivers and reservoirs likely to be affected by energy development. In the Upper MRB, the Yellowstone tributaries, the Belle Fourche River, and the Little Missouri River are especially likely to be affected by developments in Montana, Wyoming, and North Dakota. In the Upper CRB, tributaries most likely to be affected are the San Juan River and streams tributary to the Upper Colorado main stem; these will be affected by developments in Colorado, Utah, and New Mexico. Other rivers and river basins in the study area can also be affected by energy development, including the Rio Grande and the Canadian in New Mexico, the Arkansas in southeastern Colorado, the North Platte in the Lower MRB, and rivers in southern Utah which are part of the Lower CRB. However, because the greatest levels of energy development are likely to occur within the Upper MRB and Upper CRB, the following discussion will focus on these two river basins.

The total flow in these river basins and the quantities available for future uses are unclear because of data deficiencies and legal factors. However, general estimates can be made. Under the provisions of the Colorado River Compact (1922), Upper Basin states guarantee Lower Basin states 75 million acre-feet (MMAF) during each consecutive ten-year period or an average of 7.5 million acre-feet per year (MMAFY). As part of the Mexican Water Treaty (1944), the U.S. agreed to supply Mexico 1.5 MMAFY from the Colorado River. Assuming that the Upper Basin states are responsible for supplying half the treaty requirements, 8.25 MMAF must flow into the Lower Basin each year, on the average. The Upper Basin states apparently do not accept this assumption, and this adds uncertainty to quantities of water to be appropriated within each basin and within each state.

The Colorado River Compact allocations are based on an average annual virgin flow estimate of 16.2 MMAFY for the Colorado River. However, this estimate is now considered to be too high. Three of the most frequently cited average annual virgin flow estimates for the Colorado are shown in Table 2-1. Using the figures of Table 2-1 and assuming that 8.25 MMAF must be delivered to the Lower Basin each year, an estimated 5.25 to 6.65 MMAF would be available yearly to the Upper Basin, on the average.

Current average annual depletion in the Upper Basin based on 1975-76 levels of development is estimated to be 3.83 MMAFY (Colorado Dept. of Natural Resources, 1979). The breakdown of this water use is shown in Table 2-2. As indicated, approximately 56 percent is depleted by agriculture, about 20 percent is exported out of the basin, and about 19 percent is lost by

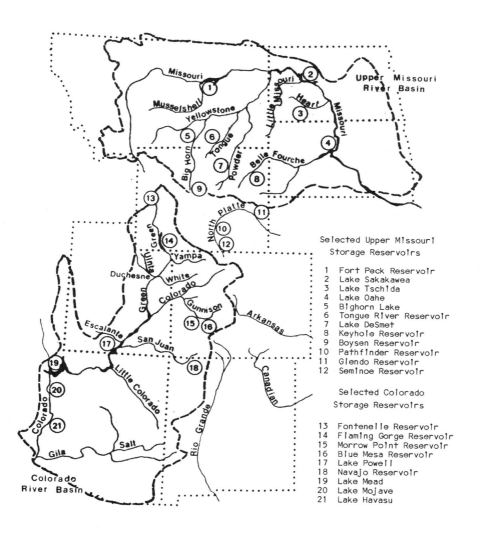

Figure 2-1:   Major Rivers and Reservoirs Likely to be
              Affected by Energy Development

TABLE 2-1:   Estimates of Average Virgin Flow in the
             Colorado River at Lees Ferry, Arizona[a]
             (MMAFY)

Source	Average Virgin Flow[b]	Time Period of Estimates
Colorado River Basin Project Act (1968)	14.9	1906-1965
Tipton and Kalmbach (1965)	13.8	1921-1964
Stockton and Jacoby (1976)	13.5	1512-1961

[a]Lees Ferry, Arizona, is the dividing point between the Upper and
Lower Colorado Basins established by the 1922 Compact.

[b]Virgin flow is the estimate of total flow had there been no
diversions or storage reservoirs in the basin.

TABLE 2-2:   Average Annual Depletions of the Upper
             Colorado River Basin Based on 1975-76
             Levels of Development
             (thousands of AFY)

Use	Amount
Agriculture	2,145
Exports	764
Evaporation from Storage Reservoirs	715
Thermal Power Plants	74
Minerals	55
Municipal and Industrial	45
Fish, Wildlife, and Recreation	33
Total	3,831

Source: Colorado, Dept. of Natural Resources, 1979.

evaporation from the storage reservoirs of the Upper CRB. Comparing depletions with the range of availability in the Upper Basin indicates that 58 to 73 percent of the available water is currently depleted.

Considerable variation exists in the average annual virgin flow in the CRB. Stockton and Jacoby (1976) estimate the average annual virgin flow at Lees Ferry to be 13.5 MMAF with a standard deviation of 3.4 MMAF. This means that in 67 percent of the years, the virgin flow would be between 10.1 and 16.9 MMAF. In drought years, flow could be much less. The year 1977 was the driest on record, having an estimated virgin flow at Lees Ferry of 5.5 MMAF (U.S., DOI, BuRec 1978a).

Protection against low-flow years is provided by a series of large storage reservoirs, most built by the U.S. Bureau of Reclamation (BuRec).[1] Total storage capacity in the ten existing BuRec reservoirs of the CRB is estimated to be about 61 MMAF. As shown in Table 2-3, from October 1976 through September 1977, actual storage capacity was about the same for the Upper and Lower Basins. About 85 percent of the total capacity is contained in Lake Mead in the Lower Basin and Lake Powell located on the southern border of the Upper Basin.

In the Upper MRB as a whole, water availability greatly exceeds any anticipated requirements over the next two decades. Major river flows based on 1970 levels of development in the Fort Union and Powder River coal regions of the Upper MRB are shown in Table 2-4. The 8.8 MMAFY in the Yellowstone contribute almost half the total flow into the Missouri above Lake Sakakawea (see Figure 2-1). Total undepleted water supply and use for 1975 in the Montana and Wyoming portions of the Upper MRB are shown in Table 2-5. Total depletions are only 16 percent of the 20 MMAFY available in Montana and 19 percent of the nearly 8 MMAFY available in Wyoming. Data on categories of depletions for the Fort Union region of the Upper MRB in North Dakota are not available.

The total average depletion in the Upper MRB is about 6.5 MMAFY including reservoir evaporation above Sioux City, Iowa (Northern Great Plains Resources Program 1974). The virgin flow at that point is approximately 28.3 MMAFY of which 19 MMAFY are estimated to be the practical limit for depletions (U.S., DOI, BuRec 1975). Hence, at present, an additional 12.5 MMAFY are apparently available for use. This compares with the 1.42 to 2.82 MMAFY currently unused in the Upper CRB.

However, water is scarce in some areas of the Upper MRB and in many cases surface water would have to be transported long distances to support energy development. For example, the coal-producing area around Gillette, Wyoming, is now water-short.[2] Further, transfers out of the Yellowstone River Basin of any surplus reservoir water may be constrained by the Yellowstone River

TABLE 2-3:   Bureau of Reclamation Reservoirs of
the Colorado River Basin
(thousands of AF)

Reservoir	Maximum Active Storage[a]	Actual Storage 1976-1977[b]
Upper Basin		
Fontenelle	345	220-300
Flaming Gorge	3,749	2,080-3,474
Blue Mesa	830	221-606
Morrow Point	117	115-115
Crystal Dam	18	0-14
Navajo	1,696	1,000-1,250
Lake Powell	25,002	16,000-19,640
Total	31,757	19,636-25,399
Lower Basin		
Lake Mead	27,377	19,600-22,000
Lake Mojave	1,810	1,460-1,794
Lake Havasu	619	540-600
Total	29,806	21,600-24,394
Upper and Lower Basin Totals	61,563	41,216-49,793

Source: U.S., DOI, BuRec 1978a.

[a]Does not include dead storage (not recoverable) of approximately 4.5 MMAF.

[b]Water year is October 1976 through September 1977.

TABLE 2-4:   Flow in Major Streams in the Fort Union and
             Powder River Coal Regions of the Upper
             Missouri River Basin[a]
             (thousands of AF)

River and Location	Average Annual Flow
Yellowstone Basin	
Clarks Fork Yellowstone	767
Wind-Bighorn (near mouth)	2,550
Tongue (near mouth)	304
Powder (near mouth)	416
Yellowstone (near Sidney)	8,800
Western Dakota Tributaries	
Little Missouri (near mouth)	390
Knife (near mouth)	118
Heart (near mouth)	154
Cannonball (near mouth)	149
Grand (near mouth)	156
Missouri River at Lake Sakakawea	16,952
Missouri River at Oahe Reservoir	18,525

Source: Northern Great Plains Resources Program 1974, p. 13.

[a]Based on 1970 levels of development.

TABLE 2-5:   Water Supply and Use in The Upper Missouri
             River Basin
             (thousands of AFY)

	Montana	Wyoming
Total Undepleted Water Supply[a]	20,141	6,884
Estimated Depletions for 1975[a]		
Irrigation	2,280	1,245
Municipal and Industrial	99	29
Minerals and Mining	10	55
Thermal Electric	1	3
Other	204	--
Reservoir Evaporation	603	172
Total Depletions	3,197	1,504

Source: U.S., DOI, BuRec 1975, pp. 299, 300, 411, 412.

[a]Water supply and depletion estimates are only for the Upper
Missouri portion of the states. The states include portions of
other river basins as well.

Compact (1950), requiring agreement among the three signatories to the compact (Wyoming, Montana, and North Dakota) (see Table 2-6). Also, there are continuing conflicts over what constitutes permissible beneficial uses along some stream segments; and many of Montana's streams have been overappropriated (U.S., DOI, BuRec 1975, pp. 298-300).

Groundwater is also available in both river basins and, in some cases, might be an alternative water source for energy development. However, it is not possible to estimate the extent of this potential because of inadequate knowledge of the location and properties of the aquifers. Groundwater quality and well yield vary significantly across the region. Site-specific studies are generally necessary to measure the depth, permeability, and water quality of aquifers before the technical feasibility and cost of using groundwater can be determined. Groundwater is currently being used extensively throughout the region, primarily for municipal, domestic, livestock, and irrigation uses. For example, it is estimated that about 133,000 AFY of groundwater are used in the Upper CRB (U.S., DOI, BuRec 1975, p. 35).

An estimated 50 to 115 MMAF are contained in shallow alluvial aquifers in the Upper CRB at a depth of less than 100 feet, of which about 15 percent is recoverable (Price and Arnow 1974). These aquifers are recharged at a rate of 4 MMAFY by infiltration and seepage from streams. However, in most of the CRB, aquifers and surface streams are hydrologically interconnected; that is, extractions of groundwater will reduce flows in surface streams. Thus, use of groundwater from such aquifers would be charged against the basin depletions of the state in which the groundwater pumping was occurring. The exception to this is deeper aquifers which are estimated to contain substantially more water than shallow aquifers.

In the Upper MRB a total of 860 MMAF is estimated to be stored in the upper 1,000 feet of rock (Missouri Basin Inter-Agency Committee 1971). But most of these aquifers are not highly productive and will be used primarily to supply water for population growth associated with energy development. Although the Madison aquifer in northern Wyoming, southern Montana, and western North Dakota has been studied as a source of water for energy development (Dinwiddie et al. 1979), its use is limited by insufficient knowledge about its water yielding properties and by its great depth in many areas of the basin; for example, it is about 7,500 feet deep near Colstrip, Montana. Alluvial aquifers and aquifers in the Fort Union Formation are presently used only for municipal and agricultural needs, although they could provide short-term supplies for energy development.

TABLE 2-6: Selected Reservoirs of the Upper Missouri
River Basin
(thousands of AF)

Stream	Reservoir	Storage			
		Inactive and Dead	Active	Flood Space	Total
Missouri	Fort Peck	4,300	10,900	3,700	18,900
	Lake Sakakawea	5,000	13,400	5,800	24,200
	Oahe	5,500	13,700	4,300	23,500
Milk	Nelson	18.7	66.8	-	85.5
Wind-Bighorn	Bull Lake	0.7	151.8	-	152.5
	Boysen	252.1	549.9	150.4	952.4
	Buffalo Bill	48.2	373.1	-	421.3
	Bighorn (Yellowtail Dam)	502.3	613.7	259.0	1,375.0
Powder	Lake De Smet	0	239.0	-	239.0
Tongue	Tongue	5.9	68.0	-	73.9
Heart	Lake Tschida (Heart Butte Dam)	6.8	69.0	150.5	226.3

Source: U.S., DOI, Water for Energy Management Team 1975.

CONSUMPTIVE WATER REQUIREMENTS FOR ENERGY

The amount of water required for future energy development in the study area is subject to considerable uncertainty, particularly regarding the total magnitude
and types of energy development that will occur (e.g.,
how large will the oil shale industry be by the year
2000?); and the consumptive water requirements for a
given energy technology. The amount of water consumed
by an energy facility depends on its type and location,
the load factor for the plant, the process design and
water management strategy for the facility, and the type
of cooling method employed (if any). Table 2-7 presents
data on consumptive water requirements for various energy technologies, both on a "standard" plant-size basis
and on a per-unit-of-energy output basis. The data are
based on the range of technologies and sites considered

TABLE 2-7:   Consumptive Water Requirements for
             Energy Facilities[a]

Facility and "Standard" Size	Water Consumption per Facility (AFY)	Water Consumption[b] per Unit of Energy Output $(gal/10^6 Btu's)$
Coal-Fired Power Plants (3,000 MWe)	23,900–29,800	124–155 (electric) 44–55 (thermal)
Coal Gasification[c] (250 MMscfd)	4,890–8,670	19–34
Coal Liquefaction[d] (100,000 bbl/day)	9,230–11,750	16–21
TOSCO II Oil Shale (100,000 bbl/day)	12,900–18,600	23–34
Modified In Situ Oil Shale (100,000 bbl/day)	7,600	14
Uranium Mine and Mills (1,000 mtpy)	270–300	.3–.33
Slurry Pipelines (25 MMtpy)	13,500[e]–18,400	16
Surface Coal Mines (25 MMtpy)	Neg.–1,240	Neg.–1
Geothermal[f] (100 MWe)	12,700–13,700	1,850–1,990 (electric) 648–697 (thermal)

$gal/10^6$ Btu's = gallons per one million   bbl/day = barrels per day
  British thermal units   mtpy = metric tons per year
MWe = megawatt-electric   MMtpy = million tons per year
MMscfd = million standard cubic feet per day   Neg. = negligible

[a]The range shown is the range of consumptive water requirements found across our data
sources and/or our six site-specific scenarios. The figures for conversion facilities include the requirements for associated mines and waste disposal. All cooling
requirements are assumed to be met by wet cooling towers. Load factors and total
annual energy output levels for each of the facilities are as follows: Power
plants--70 percent, 6.28×10^{13} Btu's (electrical), 17.0×10^{13} Btu's (thermal);
Coal gasification (Lurgi and Synthane processes)--90 percent, 8.21×10^{13} Btu's; Coal
liquefaction (Synthoil process)--90 percent, 18.4×10^{13} Btu's; TOSCO II and Modified
In Situ Oil Shale--90 percent, 17.8×10^{13} Btu's; Uranium mine and mills--100 percent, 29.9×10^{13} Btu's; Slurry pipelines--100 percent, 40.0×10^{13} Btu's; Coal
mines--100 percent, 40.0×10^{13} Btu's; Geothermal--75 percent, 2.24×10^{12} Btu's
(electric), 6.4×10^{12} Btu's (thermal).

[b]To provide a basis for comparing power plants with synthetic fuel plants, water
requirements are expressed on both an electrical output basis, and a per Btu of thermal input to the plant basis. However, comparing electric energy to energy in oil or
gas can be misleading since electricity has high quality uses not possible with oil
or gas. In the power plant, the energy produced when measured in Btu-electrical is
about three times less than the energy as measured in Btu-thermal. Neither measure
is exactly comparable to the energy content of oil or gas. A Btu-electrical is more
valuable than a Btu of oil or gas; but a Btu-thermal in the power plant case is less
valuable than oil or gas made from coal. (Also, see discussion in text.)

[c]Range shown is based on variations among sites and difference between Lurgi and
Synthane gasification processes.

(continued)

Table 2-7:   (continued)

---

[d]These data are based on the Synthoil process. Although this particular process is not likely to be commercialized, the data are representative of other types of direct coal liquefaction processes.

[e]The low end of the range is based on the coal/water ratio for the proposed ETSI coal slurry pipeline (Odasz 1980).

[f]Geothermal water use is based on projections for hot brine systems. The figures may represent an upper bound on a per-unit-of-energy basis (see text).

in our study (see White et al. 1979a; White et al. 1979b; and Gold et al. 1977 for more detailed descriptions of this data). For conversion facilities, the data are based primarily on the use of wet cooling towers, although in synthetic fuel plants some dry cooling of very high temperature gas streams is assumed. These data also assume process design optimization to reduce water consumption. The use of alternative cooling systems and process design changes to minimize water consumption in synthetic fuel plants is discussed in detail in Chapter 4.

On a "standard" plant-size basis, coal-fired power plants are the largest water consumers, requiring approximately 24,000 to 30,000 AFY for a 3,000 MWe plant. At the other extreme, uranium mines and mills require only modest amounts of water. Coal mines consume only small amounts of water for dust control, except when water is required for mined land revegetation.

For the oil shale processes, surface retorting using the TOSCO II process can consume approximately twice as much water as modified in situ techniques. Although not included in the table, water consumption for oil shale development can be reduced by about 40 percent by using a direct surface retorting process, such as Parahoe Direct, rather than indirect surface retorting processes, such as Parahoe Indirect or TOSCO II. Therefore, modified in situ processes would have water requirements similar to those for the direct surface retorting processes (Nevens et al. 1979).

The very wide range of estimates for coal gasification reflects both variations among sites and between the Lurgi and Synthane coal gasification processes. A Synthane complex consumes from 1.5 to 2.3 times more water than a Lurgi facility at the same site. In part, this is because the Lurgi process utilizes the moisture content in the coal while the Synthane process must dry the coal first.   To illustrate site variations, a Lurgi facility in the Four Corners area consumes about twice as much water as the same facility in the Northern Great Plains.   This is because New Mexico coal has lower moisture; its higher ash content requires more water for disposal; and supplemental irrigation requirements are greater to reclaim land in the Four Corners area.

Consumptive water requirements among technologies can also be compared on a per-unit-of-energy-produced basis, although such comparisons should be interpreted with caution since the energy resource inputs and product outputs are not directly comparable.   Table 2-7 presents consumptive water requirements for combustible fuel processes, based on the heating value of the output fuel.   For electricity generation from coal-fired and geothermal power plants, the high values are based on Btu output of electricity, while the low values are based on thermal input to the power plant (assuming 35 percent conversion efficiency).   Neither measure is exactly comparable to the energy content of synthetic oil or gas: a Btu-electrical is more valuable than a Btu of fuel, but a Btu-thermal input to a plant is less valuable than a Btu of oil or gas product.   However, the Btu-thermal figure for a coal-fired power plant is comparable to the figures for coal mines and slurry pipelines since both are calculated on a per-unit-of-coal-energy basis.

Given these limitations, geothermal plants are by far the largest water consumers, requiring 12 to 15 times more water than coal-fired power plants.   This is explained by the very low conversion efficiency of hot water geothermal processes resulting from the relatively low-temperature heat source, which means there is a correspondingly higher cooling requirement per unit of energy produced (see White et al. 1979a, vol. 4).   However, it should be noted that geothermal water consumption can vary greatly depending on process design.   Also, these huge consumptive water requirements may be offset by the fact that in many designs considered thus far, the cooling water will be supplied by the hot water source itself, after it has been allowed to cool in holding ponds; therefore, water from outside sources may not be required.

Coal-fired power plants are also relatively large water consumers.   For example, based on coal input, coal-fired power plants consume at least three times as much water as a slurry pipeline.   Among coal synfuels, present knowledge suggests that consumptive water requirements for various gasification facilities are larger than for direct liquefaction processes.

To illustrate the nature of the water-consuming steps, Figure 2-2 gives a breakdown of water requirements for four coal conversion facilities in Gillette,

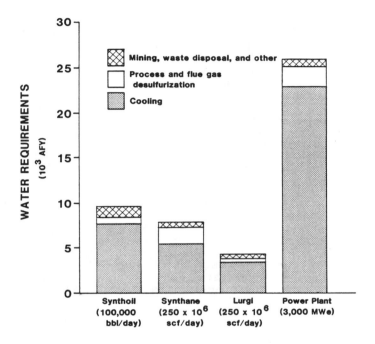

Figure 2-2:   Breakdown of Consumptive Water
              Requirements for Facilities at
              Gillette, Wyoming

10^3 AFY = 1,000 acre-feet per year
10^6 scf/day = 1,000,000 standard cubic feet per day

Source: Gold et al. 1977, pp. 192-94.

Wyoming. The breakdown is into three main groups:
cooling; process and flue gas desulfurization (FGD); and
solids disposal, mining, and other. Cooling consumes
much more water than the other uses. This is particularly true for power plants; cooling accounts for approximately 88 percent of their water consumption. The
magnitude of water consumption for FGD in power plants
is also significant, accounting for about 2,400 AFY in
the Gillette case. For all the facilities, water requirements for the associated mine, waste disposal, and
other items account for only 10 percent at most of the
total consumptive water requirements.

In order to provide perspective on total consumptive
water requirements for the region, consumptive water requirements corresponding to our Low Demand and Nominal
Demand scenarios were calculated. The results of these
calculations for the year 2000 are summarized in Table
2-8. Water consumption by energy facilities and the associated population increases range from 1.260 to 1.436
MMAFY for the Low Demand scenario to 2.719 to 3.009
MMAFY for the Nominal Demand scenario. Comparing these
consumptive water requirements with the total unused
water in the Upper Colorado and Upper Missouri River
Basins of 13.920 to 15.320 MMAFY indicates that physical availability of water should not be a problem for
energy development on a region-wide basis. That is,
energy development at these levels would at most consume
22 percent of the remaining water supplies in the two
major river basins. This is true even though the
Nominal Demand scenario represents a very massive level
of energy production.[3]

TABLE 2-8: Consumptive Water Requirements in the
Eight-State Study Area for Low and Nominal
Demand Scenarios, Year 2000

Scenario	Energy Supply (Quads)		Water Consumption (thousands of AFY)		
	National	Eight-State Study Area	Energy Facilities	Associated Population	Total
Low Demand	124.0	42.6	1,149-1,325	111	1,260-1,436
Nominal Demand	155.1	60.7	2,510-2,880	209	2,719-3,009

Quad (or Q) = 10^{15} Btu's.

However, the more important point is that analysis of water demand compared to supply on a regional basis can be very misleading. Table 2-9 shows consumptive water requirements in the year 2000 for the Upper CRB and Upper MRB and for the various types of energy facilities.[4] In the Upper MRB, energy development at the Low Demand level would account for 6.8 to 7.8 percent of the remaining unused water supplies (i.e., 0.861 to 0.981 MMAFY demand versus 12.5 MMAFY supply). In the Upper CRB, energy would consume between 10 and 24 percent of remaining supplies (i.e., 0.288 to 0.344 MMAFY versus 1.42 to 2.82 MMAFY supply). Figure 2-3 compares current (1975) uses, projected energy needs for the Low and Nominal Demand scenarios, and future nonenergy increases (estimated between 1.5 and 1.6 MMAFY) with the range of water availability estimates. By the year 2000, water consumption in the Upper CRB for the Low Demand scenario approaches the low availability estimate, while in the Nominal case demand exceeds the minimum availability estimate by more than one MMAFY. Thus, even in the more realistic Low Demand case, water demands for energy and other uses by the year 2000 will represent a large proportion of the remaining physically available supply in the Upper CRB.

The figures presented above can be compared with those from a recent study (1979) by the Colorado Department of Natural Resources (DNR). Table 2-10 presents data from this study on consumptive water requirements for their "medium" case of synthetic fuel development. Water consumption is projected to be 5.4 MMAFY. Compared to the range of availability estimates for the Upper CRB of 5.25 to 6.65 MMAFY, this indicates that demand could slightly exceed supply for the low availability estimate.[5] Note that the energy-related water consumption figures in Table 2-10 are 563 thousand AFY, almost twice those in our Low Demand scenario.

Thus, from a basin-wide perspective, water supplies physically exist in the Upper MRB to support very high levels of energy development. In the Upper CRB, water is more scarce and energy development could consume a substantial fraction of currently unused water. However, in both cases, strict reliance on basin-wide estimates to predict future water availability problems would ignore several major questions. First, even if water shortages are not of concern basin-wide, there will certainly be water availability problems in some areas. Water is relatively scarce in many energy development areas, including central and northern Wyoming, southeastern Montana, the oil shale and coal areas of western Colorado, and in the Four Corners area of New Mexico.[6] Second, there is substantial uncertainty surrounding estimates of average annual virgin flows and projections of future uses by various economic sectors,

42

TABLE 2-9: Consumptive Water Requirements for Low Demand Scenario, Year 2000[a] (thousands of AFY)

Facility and "Standard" Size	Upper Colorado River Basin			Upper Missouri River Basin			Total Water Requirements for the Region
	Water Requirements per Facility	Number of Facilities	Total Water Requirements	Water Requirements per Facility	Number of Facilities	Total Water Requirements	
Coal-Fired Power Plants (3,000 MWe)	28.5-29.8	3	85.5-89.4	23.9-26.7	14	335-374	421-463
Coal Gasification (250 MMscfd)	7.1-8.7	1	7.1-8.7	4.9-7.8	27	132-211	139-220
Coal Liquefaction (100,000 bbl/day)	11.8	0	0	9.2-10.3	2	18.4-20.6	18.4-20.6
TOSCO II Oil Shale (100,000 bbl/day)	12.9-18.6	2	25.8-37.2	12.9-18.6	0	0	25.5-37.2
Modified In Situ Oil Shale (100,000 bbl/day)	7.6	3	22.8	7.6	0	0	228
Uranium Mine and Mills (1,000 mtpy)	.3	35	1.0	.3	22	6.6	7.6
Slurry Pipelines (25 MMtpy)	19.2	1	19.2	19.2	13	250	269
Coal Mines (25 MMtpy)	0-1.24	23	0-28.5	.5	237	119	119-148
Geothermal Power Plants (100 MWe)	12.7-13.7	10	127-137	12.7-13.7	0	0	127-137
Total	-	-	288-344	-	-	861-981	1,149-1,325

[a]Based on the facility water requirements of Table 2-7, which assume wet cooling.

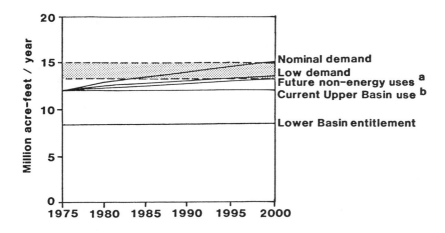

Figure 2-3:   Projections of Consumptive Water
              Requirements and Availability in the
              Colorado River (year 2000)

[a]Estimated at 1.5 to 1.6 MMAFY (U.S., DOI, Water for
Energy Management Team 1974). Since this estimate was
made, water demands have grown more slowly than forecast
and thus this estimate may be somewhat high. For example, as discussed below, a more recent study by the
Colorado Department of Natural Resources projects, in
their "medium" scenario, an increase in nonenergy uses
of 1.3 MMAFY.

[b]Estimated at 3.6 MMAFY (U.S., DOI, BuRec 1975).

including energy. Finally, any analysis based strictly
on the physical availability of water is seriously misleading; water use in the western U.S. is a sensitive
political and social issue, not simply a technical or
economic one. Important issues concern the conflict between agricultural and industrial water use, Indian water rights, water quality, such as salinity levels, and
environmental uses of water, such as maintaining minimum
instream flows. Any of these issues can affect the future use of water, including that for energy development,
as has been demonstrated in some instances by the difficulty slurry pipeline companies have obtaining water
rights. The most significant of these issues and conflicts are described in the remainder of this chapter.

TABLE 2-10: Projected Water Consumption for the
Upper Colorado River Basin, Year 2000[a]
(thousands of AFY)

Sector	Current Consumption 1975-1976	Medium Scenario 2000
Agriculture	2,145	2,736
Thermal electric	74	311
Fish and wildlife	33	74
Minerals	55	115
Municipal and industrial	45	97
Exports	764	1,149
Evaporation	528	700
Alternative energy technologies	--	252
Total	3,644	5,434

[a]This medium scenario corresponds to the study's medium "without emerging energy technology" scenario combined with the baseline "emerging energy technology" scenario.

UNCERTAINTY AND COMPLEXITY OF CURRENT WATER POLICY SYSTEMS

• *Water policy is a complex combination of state water law, federal water policy, court cases, interstate agreements, and international treaties. Although this system has dealt successfully with many difficult water problems, it also creates barriers to change, discourages diversity of water uses, and provides few incentives for efficient management and conservation of the resource.*

Because water has always been a scarce resource in the West, the history of legal and physical controls on the resource is complex and lengthy. In general, western water policy has addressed two primary concerns. The first is geographic equity: diversions and impoundments, legal doctrines, and interstate compacts were established to protect downstream users and to ensure minimal supplies to each of the states of the Colorado and Missouri river basins. The second concern is protecting the interests of irrigated agriculture, a basic industry often dependent on water supplies from distant sources. Irrigation generally accounts for 80 to 85 percent of total consumptive water use in the West (Glenn and Kauffman 1977).

Several intergovernmental arrangements regulate western water resources. As shown in Table 2-11, the waters of both the Colorado and Missouri rivers are divided by interstate compacts or agreements. One of the most important of these is the 1922 Colorado River Compact (discussed in the previous section) which established formulas for distributing water between the Upper and Lower basins. Water available to each state of the CRB is also affected by the Mexican Water Treaty (also discussed previously).

Within these international and interstate arrangements, states have primary authority for allocating water. All eight states in the study area now distribute surface waters under appropriation doctrines[7] which establish a priority system for water use based on seniority and preference clauses. Seniority is generally determined by the date of the water right application, not the date of actual diversion of water. Junior users, whether upstream or downstream, are entitled to water only after senior users have met their needs. Six of the eight states in our study area (Arizona, Colorado, Utah, Wyoming, South Dakota, and North Dakota) also have preference clauses to determine priorities for water use. During drought or other circumstances in which there is insufficient water for all uses, these clauses allow for the condemnation of water rights held by less preferred uses for allocation to higher uses, even if the preferred use has a junior right. Preference clauses usually place municipal use first, agricultural use second, and industrial use third.

An appropriation right is generally quantified by the state, and sustained only by actual and continuous "beneficial use." Defined differently by the eight states, beneficial use is a critical element of water policy, in part because withdrawal and use of water for a purpose the state considers beneficial is generally required for the water right to become perfected, or legitimate.

The beneficial use doctrine and related aspects of prior appropriation systems have created several barriers which retard changes in water use patterns and discourage both multiple uses of water and water conservation. Because states have been responsive to agricultural water interests, other values, such as protection of fish and wildlife, have achieved less legal recognition. For example, because the legal system has required that water be withdrawn from the stream in order to be considered a beneficial use, instream uses such as scenic beauty or support for aquatic ecosystems have not, until recently, been considered beneficial. As a result, many environmental groups have pursued their interests in the courts. This has often contributed to uncertainty because the time required for the litigation to be completed is long,

TABLE 2-11: Interstate Compacts and Agreements

Compact or Agreement	Lower Colorado River Basin			Upper Colorado River Basin				Upper Missouri River Basin		
	Arizona	California	Nevada	Colorado	New Mexico	Utah	Wyoming[a]	Montana	North Dakota	South Dakota
Colorado River Compact, 1922	Guarantees 7.5 MMAFY to the Lower Basin									
Boulder Canyon Project Act, 1928, and Arizona v. California, 1963[b]	2.8 MMAFY	4.4 MMAFY	0.3 MMAFY							
Upper Colorado River Basin Compact, 1948	50,000 AFY			51.75% of flow remaining after Lower Basin and Arizona	11.25% of remainder	23% of remainder	14% of remainder			
Yellowstone River Compact, 1950[c]										
Clark's Fork River							60%	40%		
Big Horn River							80%	20%		
Tongue River							40%	60%		
Powder River							42%	58%		
Belle Fourche River Compact, 1943[d]							10%			90%

TABLE 2-11: (continued)

[a]Surface waters in Wyoming are part of both the Upper CRB and the Upper MRB.

[b]Note: In Arizona v. California, the U.S. Supreme Court held that the compact-apportioned water available to the Lower Basin states was divided among them as follows: California, 4.4 MMAFY; Arizona, 2.8 MMAFY; and Nevada, 0.3 MMAFY. Although the division was not agreed to by the states, the terms of the division were finalized by the Supreme Court's decree.

[c]The compact recognized existing appropriations for beneficial uses and divided the remaining waters as shown.

[d]Recognized existing water right priorities of South Dakota and Wyoming. Remaining unappropriated water allocated as shown.

and because court rulings typically address only narrowly defined segments of the issue.

State appropriation laws also regulate how individuals may use water rights. Historical use, nonimpairment, "use or lose," and a variety of other doctrines have been developed to protect the rights of water users (White 1975). However, these doctrines have also built rigidity into the system. For example, if users have been diverting only a part of their total water rights for several years, it is not likely that they would be able to begin diverting all of their potential allocation if downstream users had become dependent on the unused part of their allocation. This aspect of the prior appropriation system, sometimes referred to as "historical use," creates an incentive for users to "use or lose" all of their water rights.

The rigidity of the appropriation system is one of the largest barriers to water conservation, since farmers risk losing the right to water which they might save, for example, by using more efficient conveyance or irrigation systems. This occurred in the case of Salt River Valley Water Users Association v. Kovacovich (1966) in which the Arizona Supreme Court denied the right of a water user to apply water "salvaged" by conservation to lands other than those to which the water was originally appurtenant or attached.

State appropriation systems also create barriers to resource management. This is because the system is primarily designed to codify legal water rights and uses, rather than serve as a tool for managing the resource. For example, although states generally record the date, priority, and use for which a water right is granted, little if any information is recorded about actual use practices. In part, this uncertainty about actual water use is caused by the number of water applications which are not "perfected"--that is, even though the permit has been granted and a priority assigned, it may remain unexercised for many years. For example, in Colorado there are conditional decrees for water rights which are over thirty years old. The effect, as noted by Harris Sherman, the former Executive Director of the Colorado DNR, is that "most streams in the state are over appropriated due to water rights filings" (Shupp 1980, p. 54). Some filings and conditional decrees for water rights have been obtained by energy developers (see Table 2-12), but the relatively late dates of application make these water rights junior to many of the water rights held by the agricultural sector.

Further, little data exists regarding return flows; typically, data gathering in this respect begins only when users claim their rights have been illegally impaired by another use. Without data on actual use practices and return flows, determination of the actual

TABLE 2-12:   Water Right Applications for Colorado
Oil Shale Area[a]

Applicant and Number of Applications	Range of Dates of Appropriation	Quantity of Water Claimed AFY	Status[b]
Union Oil Company (1)	6/49	85,770	CD
Cities Service Oil Co. (1)	8/51	72,380	CD
Chevron Oil Co. (3)	11/51 to 6/53	181,860	CD
Getty Oil Co. (1)	11/51	40,530	CD
Colony Development Operations[c] (4)	6/53 to 1/55	171,274	CD
Atlantic Richfield Co. (1)	11/56	23,885	CD
Mobil Oil Corp. (1)	6/61	36,190	CD
Humble Oil & Refining Co. (2)	12/63 to 10/64	217,140	F-CD
Sohio Petroleum Co. (2)	2/65	72,380	CD
White River Resources, Inc. (1)	8/66	72,380	F
Industrial Resources, Inc. (1)	10/66	72,380	F
The Oil Shale Corp. (1)	2/67	39,809	F
Superior Oil Co. (1)	5/68	17,370	F

Source: U.S., DOI 1973, p. II-30.

[a]All data obtained from water filings made with Colorado State
Engineer. Virtually all of these direct flow applications are
accompanied by separate appropriately sized storage reservoir
applications.

[b]CD = Conditional Decree has been awarded by Colorado State Court.
 F = Filed only--no conditional decree yet awarded.

[c]Partners in Colony are Atlantic Richfield Co., The Oil Shale
Corporation, Sohio Petroleum Co., and Cleveland Cliffs Mining Co.

consumptive use of water rights as well as administration of transfers is impeded.

The prior appropriation legal system rigidly defines
how water can be used and often makes transfers of water
rights for other uses difficult (Khoshakhlagh 1977).
In New Mexico, for example, if a farmer wants to take
some land and water out of production and sell or rent
the water to another user, several difficulties must be
overcome. The burden of proof for establishing nonimpairment to other water rights holders is borne by the
transferee, and costs for meeting this requirement can
be high. An application for transfer is typically accompanied by surveys, maps, plans, and specifications
that must be filed with the state engineer. If the application complies with relevant statutes and regulations, the state engineer approves the application. In
most cases, the applicant is then required to publish a
notice of his intention to change a water right.
Parties who believe they may be affected by the change
of right may protest and in some cases can prevent the
change. If no party protests, the application may be
concluded. If protests are filed, the state engineer
may convene a hearing. Parties in disagreement with the
state engineer's findings may then appeal to the district
court. Often, resolution of the issue involves a lengthy
court proceeding (Khoshakhlagh 1977).

Some transfers are occurring nonetheless, and the
impetus for transfers has increased with energy development. For example, in Lynndyl, Utah, a local power consortium planning to build a 3,000 megawatt coal-fired
electric generating plant has reached agreement with
four irrigation districts in the area to purchase 45,000
AFY of water from a water transporting, or ditch, company at a cost of $1,750 for the right to each AFY
(Willardson 1979). However, there are indications in
the West that states are beginning to use their water
law as a device to control and restrict energy development (Trelease 1979). This trend, along with the procedural difficulties mentioned above, may restrict the
quantities of water available to energy industries and
limit the flexibility of choice by current water rights
holders.

INCREASED DEMANDS FOR WATER

- *The manner in which water is used for agriculture under
  the current system has come under increasing criticism,
  including charges that much irrigated agriculture wastes
  water, is uneconomical in the sense that agricultural
  crops do not return the value of the water used to produce
  them, and are environmentally damaging. Although these*

*criticisms raise important questions, the fundamental
change occurring is that diverse users are requesting a
larger part of a scarce resource. In the future, these
demands are likely to increase conflicts among potential
users throughout the two major river basins in the study
area, create water shortages in some locales, and increase uncertainty about energy resource development.*

Energy Needs

Water availability in the region and water requirements for energy have been discussed earlier in this
chapter. In our study area the most troublesome locations appear to be in central Wyoming, along the tributaries of the Yellowstone River in northern Wyoming and
southern Montana, in western Colorado along the Colorado
main stem, and in the San Juan River Basin. However,
energy development will continue to contribute to conflicts in other areas because competing water users will
be threatened.

One of the important questions in this regard is
whether or not energy industries will be able to acquire
water expeditiously[8] and, if so, whether or not it will
be at the expense of irrigated agriculture. In some
cases, state governments have tried to ensure adequate
supplies for resource development. For example, to spur
development of the Uinta Basin, the Utah water code was
revised in 1976 to allow the state engineer to approve
applications to appropriate water for industry, power,
mining, or manufacturing (Laws of Utah 1976). Other
states, particularly Colorado, Wyoming, and Montana, are
apparently much less willing to favor water for energy
development over agriculture. For example, congressmen
from Colorado have recently attempted to make synthetic
fuels development contingent on funding for new water
resource projects (see box, "Tying Synfuels to Water").
Without special projects it may be more likely in these
states for energy developers to buy rights from current
users, primarily irrigators.

How many water transfers will occur, whether they
will provide enough water for energy, and the consequences for the West are unclear. Energy resource developers
can generally pay substantially more for water than can
irrigators. For example, although the economic value of
water depends on many factors, one source estimates that
water for agriculture is usually profitable at costs up
to about $25 per AF, but water for energy can be econom-
ic at $200 per AF or more (Andersen and Keith 1977, p.
161). This economic situation has led some observers to
suggest that water for energy will not be a problem--

```
┌─────────────────────────────────────────────────────────┐
│                  TYING SYNFUELS TO WATER                  │
│                                                           │
│     During the spring, 1980, debate over federal subsidies │
│  for water project development, Colorado Senator Gary Hart │
│  warned that "...you're not going to have those synfuels  │
│  plants without new water projects in Colorado." However, │
│  Hart was opposed by other senators, who indicated that the│
│  eight years required for oil shale production start-up made│
│  immediate funding of water projects unnecessary.        │
│                                           --Larsen 1980.  │
│                                                           │
└─────────────────────────────────────────────────────────┘
```

energy producers will simply buy up water rights currently used for agriculture if other sources do not
exist.

However, several barriers exist to the large-scale
transfers of water from agriculture to industry. Among
these are the legal restrictions of state appropriation
systems (previously discussed). In addition, although
some farmers would benefit economically from water transfers, it is clear that many farmers are opposed to
large-scale transfers to industry. In some cases, farms
have been worked by a family for generations. Throughout the West, farmers' groups have organized to protect
their interests. For example, the Northern Plains Resource Council (NPRC) represents farmers, ranchers, and
citizens concerned with coal development in the northern
plains. In addition to opposing water appropriations
for power plants in Montana, NPRC supports agriculture
expansion in the Yellowstone Basin (The Plains Truth
1979, p. 5). John Stencel, President of the Rocky Mountain Farmers Union, has expressed a similar opinion
regarding competition for water between industry and
agriculture:

> We urge adoption of legislation...to prevent
> future power and energy plants from consuming
> water to the detriment of agriculture...we also
> feel that the sale of adjudicated irrigation
> water rights should be limited to agricultural
> uses (U.S., Congress, Senate Committee on Energy
> and Natural Resources 1978).

It is expected that some farmers will be willing to
sell their water rights to energy developers and the
prospect of conversion from agricultural to industrial
water uses is troublesome to many westerners. These
concerns, along with the desire of western states to

control the development of their resources, have led some states to use their water law as a device to slow or prevent energy development (Trelease 1979). This politicalization of western water law is most apparent concerning coal slurry pipelines. For example, Montana has declared that "a use of water for slurry to export coal from Montana is not a beneficial use" (Mont. Code Ann. 1978a). Also, Wyoming has passed a law preventing the export of ground or surface water for use outside the state without prior approval of the legislature (Wyo. Stats. Ann. 1977).

While the main thrust of these actions has been towards coal slurry pipelines, the trend towards controlling other types of energy development and protecting agriculture is growing. South Dakota has passed a law which requires all applications to appropriate more than 10,000 AFY to be submitted to the legislature for approval (S. Dak. Cod. L. Ann. 1979). Montana has placed a prohibition on all transfers of water rights from agriculture to industry in excess of 15 cubic feet per second (cfs) (10,860 AFY) (Mont. Code Ann. 1978b). As noted by Trelease, this unwillingness on the part of western states to switch from "irrigation law" to "energy law" may invite the increased federal intervention westerners fear (Trelease 1979, p. 19).

Further, in some cases energy-related water needs may require the water rights belonging to several farms in the vicinity of the energy resource. As described previously, some coal conversion facilities have large consumptive water requirements; a 3,000 MWe power plant (wet-cooled) consumes about 24,000 to 30,000 AFY and a 1,000,000 bbl/day coal liquefaction plant (wet-cooled) consumes about 9,000 to 12,000 AFY. In some cases, these requirements may exceed the total water consumed in an entire irrigation district or reclamation project. For example, the Colbran reclamation project irrigates nearly 20,000 acres of land along the Colorado main stem in western Colorado. It is estimated that farms benefiting from the project consume about 20,000 AFY, substantially less than the consumptive water requirements of a 3,000 MWe power plant (U.S., DOI, BuRec and BIA 1978).

Although slurry pipelines require less water on a per-unit-of-coal basis than many other types of energy technologies studied, prospective coal slurry pipelines are an increasing concern in the West, particularly in Colorado and Wyoming. For example, Houston Natural Gas Company and Rio Grande Industries Corporation of Colorado have proposed a slurry pipeline from Walsenburg, Colorado, to Texas which would carry 15 million tons of Colorado coal per year and use about 10,000 AF of water per year. However, Colorado farmers, ranchers, legislators, environmentalists, and railroad companies have opposed the development of this line (Davidson 1977). Railroad representatives, who say they operate at 30 percent capacity, have aligned with environmentalists in an attempt to prevent the energy companies from acquiring rights of eminent domain from states.

Water requirements for this slurry pipeline have also concerned Colorado legislators, some of whom proposed legislation which would allow the line to run through the state but would prohibit the use of Colorado water. This was done even though this pipeline would apparently produce $2 million to $3 million per year in state taxes. The issue was also being addressed in the courts, where the pipeline company had been asked to prove that its plan to pump groundwater would not injure water supplies reserved for present users. The possibility of lowered water tables, coupled with the fact that the pipeline companies have already purchased water rights from some area farmers, apparently has infuriated other farmers (Strain 1977).

Environmental Needs

As mentioned previously, environmental demands for water also contribute to the growing complexity of the water policy system. In some states, such as Montana, instream environmental uses recently have been legally recognized as being "beneficial." In addition, suits have been filed in each of the eight states to preserve scenic beauty, natural habitat, and recreational uses. For example, the Colorado Water Conservation Board, representing the Colorado Division of Wildlife, sued the Colorado River Conservation District, which represents water-rights holders. The issue was preservation of minimum stream flows in the Crystal River, near Carbondale, Colorado, in order to protect the natural environment, including ensuring that enough water would be available for trout (Saile 1977). This issue involved a confrontation among environmental, agricultural, and commercial power interests. It is also indicative of litigation over many streams of the West which, like the Crystal River, are over-allocated.

Environmental values articulated in the Wild and Scenic Rivers Act (1968) also make new demands on scarce resources. In the eight-state study area, parts of twelve rivers in the Colorado Basin are being considered for inclusion in the Wild and Scenic Rivers System. Those rivers most likely to have impact on energy development are the Escalante, White, Green, and the Colorado main stem between the confluences with the Gunnison and with the Dolores. The act may require the preservation of stream flows to protect instream uses. This can affect water available for energy development and other

PROJECT THREATENED

The proposed Juniper Cross Mountain Project, which would store up to one million acre-feet of water in reservoirs on the Yampa and Little Snake rivers near Craig, Colorado, may be blocked by an endangered species, the squaw fish. Because the Yampa has been identified as prime habitat of the squaw fish, a construction permit will not be given until the potential danger to the fish is determined.   --Denver Post 1978

uses and prevent impoundments or any water resources projects that would alter the nature of flow of a wild or scenic river. However state water administrators generally do not believe that this act will reduce water availability for appropriation.

Environmental interests often come into direct conflict with agricultural interests in regard to water quality. The basic question relates to irrigation practices: most agricultural lands in the West are irrigated by flooding, which requires diversion of water and subsequent return flows that add salts to the river system. This irrigation practice may result in higher than necessary diversions; one estimate suggests that at least twice as much water is typically diverted as is needed for maximum plant growth (Utah State Univ., Utah Water Research Lab. 1975, p. 60). However, this diverted water is not necessarily consumptively used and most of it finds its way back into the stream (see Chapter 4).

Many environmentalists prefer irrigation systems which would leave more water instream and, thus, improve water quality by reducing salinity caused by irrigation runoff (see Chapter 4). In fact, salinity problems have already created political conflicts (see box). Salinity may also affect energy development; because land disruptions can increase salinity loading and withdrawals for energy generally increase salinity concentration levels downstream, stricter salinity standards could make it more difficult for energy developers to receive the necessary permits. However, irrigation runoff is clearly the largest man-made source of salinity problems and this represents a dilemma to environmental interests. Because runoff creates marshes and wetlands, which can be scenic and can provide wildlife habitat, many wildlife oriented environmentalists are reluctant to advocate other types of irrigation systems.

EDF SUES FOR SALINITY CONTROL

The Environmental Defense Fund (EDF) is suing EPA to require stricter control in the Colorado River Basin. An EDF spokesman, calling salinity the most serious water quality problem in the basin, estimated the salinity cost for agricultural, municipal, and industrial users to be over $50 million annually. EDF will try to get EPA to enforce existing laws and force states to meet control deadlines. Another EDF spokesman called for the creation of a system of "salinity rights" that could be treated like current water rights in order to regulate the quality of water returned to surface waters. EDF lost in D.C. District Court in October 1979 and appealed the ruling to the D.C. Circuit Court of Appeals. The Circuit Court affirmed the lower court's decision in April 1981. As of this writing (7/81), EDF has not decided on further action.
--Gill 1977; Denver Post 1977a; Pring 1980; EDF 1981.

RESERVED WATER RIGHTS

The reserved rights doctrine may entitle Indians and the federal government to large quantities of surface water and groundwater. Indian water rights already bring Indians into direct political conflict with state governments, and jurisdictional questions about the responsibility for settling Indian claims to water add to the uncertainties about water availability. If Indian claims to water are denied, the development of Indian lands may be restricted and damage to Indian values and culture may occur.

Reserved rights recognize that when the U.S. establishes a federal reservation such as a national park, military installation, or Indian reservation, a sufficient quantity of unappropriated water is reserved to accomplish the purposes for which that particular land was reserved. The reserved rights doctrine has been defined in the courts to hold that federal reserved rights are not subject to state appropriation laws and that reserved water rights are not lost if they are not used.[9] This doctrine is significant because the federal government and Indian tribes own large areas of western land (see Chapter 1)--about 70 percent of the land in the CRB.

An important water rights issue concerns the potentially large quantities of water at stake. As shown in Figure 2-4, a considerable portion of western surface water flows through or borders on Indian reservations. Under the reserved rights doctrine, some Indian tribes have claimed rights to large quantities of water. If these claims are held to be prior and paramount, many existing allocations and appropriations could be affected. In one case, the Supreme Court has held that Indian water rights were included in interstate compacts (<u>Arizona</u> <u>v.</u> <u>California</u> 1963). If this interpretation were applied generally, each state would have to absorb any changes resulting from the exercise of Indian water rights within the state.

Federal reserved rights may also involve large quantities of water, and like Indian rights, the amounts of water necessary to fulfill the purposes for which reservations were created generally have not been quantified. However, a recent case (<u>U.S.</u> <u>v.</u> <u>New</u> <u>Mexico</u> 1978) indicates that the court will take a narrow view of the expressed intent of Congress in establishing reservations. Such an "expressed intent" criterion has not been applied to Indian reserved rights (Boris and Krutilla 1980, p. 53).

<u>Arizona</u> <u>v.</u> <u>California</u> does little to set a clear legal precedent or to contribute to the quantification of Indian or federal water rights because the amounts and purposes of water usage on the reservation remain ambiguous or undefined. First, the court established a "practicably irrigable acreage" criterion without a clear interpretation of what constitutes "practicably irrigable" land. It is likely that such a criterion could vary substantially over time and from site to site (U.S., GAO, Comptroller General 1979, p. 86). Secondly, the issue of whether nonagricultural uses of water are reserved is still unsettled. While the Supreme Court did not address this question, the court did appoint a Special Master who has noted that the decision rendered in <u>Arizona</u> <u>v.</u> <u>California</u> does not necessarily preclude nonagricultural uses of reserved rights (Boris and Krutilla 1980, p. 53).

In the <u>Winters</u> case, the Court recognized the possibility of future Indian involvement in both agriculture and "the arts of civilization." And some writers have suggested that this phrase could be interpreted to "include industry as well as agriculture" (Boris and Krutilla 1980, pp. 53-54). The Indian viewpoint generally is that water rights should be interpreted broadly to give the tribes maximum flexibility in using their water resources. To this end, Indians have claimed water rights under an expansive criterion which encompasses multiple uses. For example, in <u>U.S.</u> <u>v.</u> <u>Tongue</u> <u>River</u> <u>Users</u> <u>Association</u> <u>et</u> <u>al</u>. (1975), the Northern

58

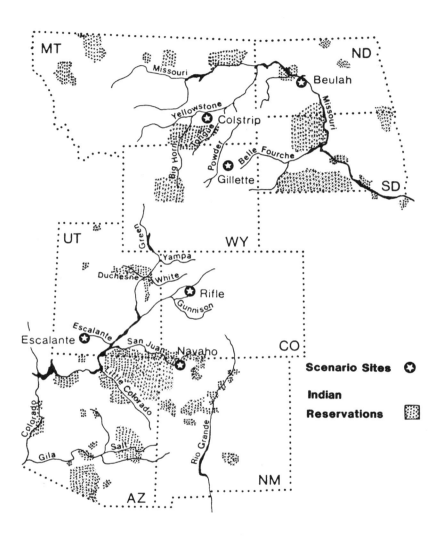

Figure 2-4: Western Rivers in Relation to Indian
Reservations

Cheyennes included environmental and aesthetic purposes
in their claim of beneficial uses of water. In a pending case that concerns the Southern Ute Indian Reservation in Colorado, the federal government has claimed for
the Indians sufficient water to bring coal to a market-
able condition (Boris and Krutilla 1980, p. 57).

Both the states and Indian tribes seek to have such
claims adjudicated in the court that is more favorable
to their case. Western states would like to quantify
Indian claims in state courts because they feel they
would have the advantage. By the same reasoning, the
Indian tribes have consistently initiated suits in federal district courts because they view state courts as
unfriendly. This continuing struggle between the states
and the Indian tribes must be seen against the background of the legal systems of the respective western
states which generally exclude Indian water rights from
state regulation or control. The enabling acts and
state constitutions of most western states contain provisions wherein states "disclaim all right and title to
Indian land and agree that such land shall remain subject to the disposition and absolute jurisdiction of Congress" (Thorpe 1977, p. 198). Despite these provisions,
however, suits for Indian water rights have in the past
been brought in state courts under the McCarran Amendment
(1952) which waives the right of sovereign immunity of
the federal government in cases concerning reserved water
rights. This amendment established concurrent state-
federal court jurisdiction over reserved rights and
opened the way for states to include reserved rights in
general adjudication procedures (Kleppinger 1977).

Concurrent jurisdiction provided by the McCarran
Amendment was extended by the Supreme Court in Colorado
River Water Conservation District v. United States
(1976). In this case, the Supreme Court indicated a
preference for reserved rights adjudication by state
courts if the state had adopted a comprehensive adjudication program (Gaufin 1977; Kleppinger 1977). In the
eight-state study area Wyoming, Colorado, and New Mexico
have adopted general adjudication programs.

The political impact of the Colorado River case and
the McCarran Amendment has been to heighten the political conflict between the federal government and the
states and, therefore, to increase the uncertainty over
water availability. As characterized by one observer,
the state, Indians, and the federal government are in a
race for the courthouse every time a claim for reserved
water rights arises. This contest has a payoff for the
winning side because a state court will not hear a case
that is pending in federal court, and vice versa. Generally, however, it is unclear in what circumstances
federal jurisdiction takes priority over state

jurisdiction, and these circumstances may have to be determined on a case-by-case basis.

INTER- AND INTRA-GOVERNMENTAL DISPUTES

Disputes between various interests have arisen over western water management, including conflicts between states of the Upper and Lower CRB, state and federal governments, and Congress and the President. These disputes are indicative of the increasing seriousness of water quality and availability problems in the West, including whether or not western water will be used to help meet the nation's energy needs. Further, they suggest that new institutional mechanisms will be necessary in order to cope with the increasing demands for multiple uses of scarce western water resources.

Disagreements over water rights, multiple uses, quality, and development have raised serious questions concerning the roles and responsibilities that various governmental units should have in managing western water resources. These disputes involve conflicts among and within states, among agencies and branches of the federal government, and between states and the federal government. These conflicts have increased the uncertainty over water availability and have made cooperative water management efforts even more difficult than they would have otherwise been.

Regarding interstate relations, the Upper and Lower CRB states have been working closely together to overcome the problems that have developed in the basin regarding salinity and pollution control burdens. However, the states have not been able to resolve their differing positions on water allocations. For instance, disagreement exists over whether Upper or Lower Basin states are responsible for providing the 1.5 MMAFY to Mexico required by the Mexican Water Treaty. Some policymakers in the area have warned that such conflicts threaten the fragile coalition of states in the CRB (Engdahl 1979).

Conflicts among branches and agencies within the federal government have added another element of uncertainty in the western water policymaking system. In part, these conflicts reflect the diversity of federal interests in the region and the fragmented relationship among the several executive-level agencies which have responsibility. These interests include encouraging energy development, preserving environmental quality, subsidizing a large fraction of the agricultural development, and regulating water quality. As the discussion in this chapter has shown, it has been increasingly difficult, if not impossible, to balance these competing interests. Thus, although western state water management policies can be criticized, it is also accurate that federal policies affecting water are disparate and conflicting.

Conflict within the federal government has frequently included the question of whether the executive or legislative branch should play the primary role in formulating federal water policies. For example, President Carter's water project "hit list," announced in 1977, attempted to reduce the number of federally-funded water projects, many of which were western. This plan was linked to a plan to provide an independent review of all new water project proposals by the Water Resources Council (WRC). Cancellation of water projects was only partially successful and the independent review proposal failed in Congress, in large part because many legislators felt that Congress was traditionally the driving force in public works policy.

Conflicts within the federal government are directly related to state-federal disputes. In fact, perhaps the most central issue in western water policy is that of individual state responsibility as opposed to a stronger federal role. The Carter Administration actively pursued a stronger federal role, in part to help further domestic energy goals, but also to improve management and efficient use of the resource (Kirschten 1977a). Beginning with Carter's water policy proposals in 1977, several proposals were made which threatened state authority in water policy. These included proposals to: (1) decrease funding for water development projects; (2) require states to share at least 10 percent of project costs; (3) enforce the "ability to pay" doctrine for recipients of water from federal projects so that a higher percentage of federal costs can be recovered; and (4) enforce a 1902 law (Reclamation Act) which limits to 160 acres the amount of land which can be irrigated from federal reclamation projects.

The impression created by these proposals on the western states was clear; it was viewed as an attempt to change the entire nature of a policymaking system which had been in use for decades. Secretary of Interior Andrus confirmed the change in direction in 1977:

The day of considering money to be the only solution to water problems is over. We want results not in the form of more dams and canals and the like, but in the form of more rational use of this very precious resource (Kirschten 1977a, p. 900).

Western states reacted sharply to these proposals. Reacting to the 160-acre limitation, Wyoming Governor Ed Herschler said the proposal would..."upset the entire western agriculture [sic]" and called for equivalency standards which would allow ranchers in arid regions to irrigate more lands than those in humid areas (Denver Post 1977b; Kirschten 1978). State reactions generally were successful in stopping, delaying, or altering most of the Carter Administration's proposals. The cost-sharing and ability-to-pay plans were altered significantly or dropped by the Administration. However, four reclamation projects to have been located in the study area were ultimately cancelled by Congress.[10]

The issue of federal control over western water re-surfaced with the issuance of a legal opinion by the Department of Interior's (DOI) Solicitor Krulitz on federal water rights. This opinion contends that:

> Since the federal government has never granted away its right to make use of unappropriated water on federal lands, ...the United States has re-tained its power to vest in itself water rights in unappropriated water and it may exercise such power independent of substantive state law (U.S., DOI, Office of the Solicitor 1979, p. 11).

Western governors were antagonized by this opinion which was regarded as an attempt to circumvent recent Supreme Court decisions declaring the federal government must comply with state laws in appropriating water, except in the case of federal reserved rights (U.S. v. New Mexico, 438 U.S. 696, 1978; California v. U.S., 438 U.S. 645, 1978). As stated by Governor Matheson of Utah:

> In his assertions concerning nonreserved federal rights, the Solicitor's opinion must at best be seen as espousing the kind of hypothetical and theoretical claims which the President rejected. At worst, it is flatly contemptuous of the Supreme Court's recent decisions on the issue (Western States Water Council 1979, p. 2).

The nature of state-federal conflicts could change with the Reagan Administration which favors regional and decentralized approaches to water management. However, many of these disputes are likely to continue for the forseeable future. It is unclear how much they will influence energy development in the West. However, they are indicative of the increasingly serious problem of water availability and water quality. The management of water resources has always been essentially a state prerogative, yet the implications of various water uses create interstate, national, and even international issues. This fact raises the question of whether new institutional mechanisms may be required to cope with the variety of demands being made on scarce western water.

SUMMARY

How much energy development will impact other water users and how much water availability will affect energy development are questions to which there are no simple answers. Our analysis and several other recent studies which examined the total amount of unappropriated water compared to the amount of water required by energy development have concluded that sufficient water physically exists on a region-wide basis to support a significant increase in western energy production without a dramatic impact on existing users, primarily agriculture (see, for example, U.S, GAO, Comptroller General 1980 and Colorado, Dept. of Natural Resources 1979). However, any analysis based strictly on the physical availability of water is incomplete. Energy development is only one of several unsatisfied demands for western water. Instream values, Indian reserved rights, federal reserved rights, possible federal nonreserved rights, expanding agriculture, and salinity control are some of the interests also demanding recognition.

The importance of water has always been appreciated in the West, but new and larger demands on the region's existing resources increase the visibility and significance of water policies. In addition to raising questions about the adequacy of resources and the distribution of limited supplies, this situation also has caused the legal and administrative system for managing water resources to come under close scrutiny. For example, the law has been criticized for often not recognizing instream values, for promoting the wasteful use of water, and for ignoring the connections between water quality and water quantity. And, there is a trend developing in western states to "politicize" water law, i.e., to use water law to protect agriculture and control the pace and location of energy resource development.

At a broad level, the issues concern the adequacy of the existing institutional system to cope with new demands both for limited water resources and for participation in the policymaking processes. Institutional innovations have occurred in the past in response to changing conditions, and it is clear that further innovations will be required as problems become regional and begin to influence national goals, such as decreased dependence on foreign energy sources. The courts have played a major policymaking role, yet they characteristically operate slowly and provide piecemeal, localized, and at best, short-term resolutions to problems. The mechanisms for resolving the issues of water availability and increased energy development will need to allow consideration of more comprehensive approaches to the management of water resources, accommodate diverse interests and values and facilitate compromises before, rather than after, large-scale impacts occur.

NOTES

[1]The name of the Bureau of Reclamation was changed to the Water and Power Resources Service in late 1979. On May 20, 1981, Secretary of the Interior James Watt restored the original name.

[2]The municipality of Gillette has inadequate water sources, storage, and distribution facilities (Gillette Municipal Government, 1979). A power plant in the Gillette area utilizes dry-cooling due to water scarcity.

[3]The Nominal Demand scenario was based on a national energy production level in the year 2000 of 155.1 Quads, with 60.7 Quads coming from the eight-state study area. Because of economic and environmental factors and the strong emphasis on energy conservation it is virtually certain that this level of development will not occur by the year 2000. This level of development may not be unreasonable at some longer term date, say 2020.

[4]This breakdown between the Upper CRB and Upper MRB is approximate. The Stanford Research Institute (SRI) model used to generate the scenarios subdivided the study area into two regions: Rocky Mountains (Arizona, New Mexico, Colorado, and Utah) and Powder River (Wyoming, Montana, and the Dakotas). Table 2-9 is based on equating the Rocky Mountain and Powder River regions with the Upper CRB and Upper MRB, respectively, which is approximately, but not exactly, correct. In fact, there will be some facilities located in other river basins of the study area (e.g., along the North Platte River in Wyoming, which is part of the Lower MRB).

[5]The Colorado DNR study did not make this comparison, but rather evaluated the effects of depletion level on the hydrology of the river basin using a computer model to evaluate a range of factors and assumptions. They concluded for this medium scenario that depletions "...are about as large as could be sustained..." (Colorado DNR 1979, p. 7-25).

[6]For an analysis of the impact of increased depletions on instream values in New Mexico, see Gilliland (1981).

[7]South Dakota has a mixed riparian-prior appropriation system; however, all new water rights fall under the appropriation system.

[8]At a 12 percent inflation rate, a one-year delay could cost $120 million for a one billion dollar project, in addition to interest on any investments already made.

[9]Specific extensions of the reservation doctrine are the product of over 70 years of refinement of the Winters ruling (Winters v. U.S. 1908). This doctrine also has been confirmed in Arizona v. California (1963).

[10]The four projects eliminated were the following: Fruitland Mesa Dam, Colorado; Lake Oahe Diversion Project, South Dakota; Savery and Pot Hook Dams, Colorado and Wyoming; and Narrows Unit Dam, Colorado (CQ Weekly 1977).

3
Water Quality:
Impacts and Issues

INTRODUCTION

Water quality has long been an issue in the West. Traditionally, attention has been centered on pollution from agricultural runoff and from natural sources. This concern has intensified recently, in part because the quality of water is closely related to its availability: as water consumption increases, the quality can deteriorate because existing pollutants become more concentrated. In addition, conflicts over water quality can be traced to increased demands for water for municipal use, increased awareness of the environmental impact of pollution, and the potential threat to water quality from the development of energy resources. These concerns over water quality have created intergovernmental conflicts as well as conflicts among water users. For example, over the past several years disputes have arisen among the federal government, the states of the Colorado River Basin (CRB) and environmental interest groups over appropriate policies and regulations for dealing with salinity problems in the Colorado River.

Energy developmemt activities will increase conflicts over water quality for several reasons:

- Energy extraction and conversion processes can cause pollution of both surface water and groundwater;

- Consumptive use of water by energy facilities will generally increase salinity concentrations downstream by reducing dilution; and

- Rapid population increases associated with energy development can overload the sewage treatment capacity of small communities.

The Clean Water Act (CWA) (1977) limits the direct discharge of industrial pollutants into surface waters. Nevertheless, energy extraction and conversion can disrupt land, resulting in increased erosion and damage to aquifers. In addition, seepage or leaching of contaminants from the disposal of liquid and solid wastes can damage water quality. The second problem area is related to the effects of water consumption and not to the discharge of pollutants. In many rivers in the West, salinity concentrations increase downstream. Therefore, the consumption of relatively fresh water upstream reduces the water available for dilution, further increasing salinity concentration levels. While salinity is a concern throughout the western region, the issue is especially important in the CRB. The third problem area is associated with rapid population increases, leading to inadequately treated or untreated municipal wastes and degradation of surface streams.

This chapter addresses each of these three water quality issues. Each section will include an assessment of the impact or potential impact that energy development poses, followed by a summary description of the current legal and institutional system for dealing with that particular problem area. Policy options for mitigating these problems are the subject of Chapter 6.

POLLUTION FROM ENERGY FACILITIES

Summary: Current federal and state regulations essentially prohibit the direct discharge of pollutants from energy facilities into surface waters. Nevertheless, energy production and conversion processes can pollute both surface water and groundwater either through the generation and disposal of waste products or through the disruption of land and the contamination of aquifers during mining and *in situ* recovery operations. Regarding aquifer contamination, information on the extent of the potential impacts and control technologies is inadequate. Control of seepage and runoff from waste disposal constitutes a serious problem, especially after a facility has shut down. Given current federal and state regulations and the level of uncertainty, existing environmental regulations are inadequate to ensure that long-term or irreversible damage to surface and groundwater quality does not occur.

The energy production and conversion technologies considered in this study pose a threat to water quality for two reasons. First, land disturbance from resource extraction activities (such as coal, uranium, or oil shale mining; oil or gas production; geothermal energy production; and in situ oil shale recovery) will disrupt groundwater aquifers and can introduce contaminants into the aquifers. Surface disturbances, such as surface coal mining, road construction, and the construction/operation of conversion facilities, increase erosion and concentrate dissolved solids (salinity) and other pollutants. With mining, groundwater pollution is primarily the result of changes in percolation or movement of groundwater through newly exposed soluble or toxic materials. For the recovery of uranium or oil shale, new hazardous materials may be introduced into the groundwater and eventually find their way into surface waters.

At each of the six hypothetical scenarios considered (White et al. 1979b) some disruption and contamination of the aquifers and nearby surface waters was projected to occur. For example, at Kaiparowitz (Utah) and Rifle (Colorado), underground mining of coal and oil shale was projected to intersect aquifers contained in the resource formation and to result in the potential degradation of fresh aquifers by saline aquifers. In other locations, such as Gillette (Wyoming) and Colstrip (Montana), the leaching of contaminants into groundwaters during surface mining of coal was projected. Solutional mining of uranium as may occur in the Gillette area could result in the leaching of pollutants into nearby aquifers. At all locations vegetation removal and soil disturbance during mining and construction activities could result in increased erosion and sediment loading of nearby streams. There have been problems in aquifer disruption, for example, in the Grants Mineral Belt area of New Mexico (New Mexico Environmental Improvement Agency 1977), in eastern Utah and western Colorado in areas of coal, oil shale and petroleum development (Price and Arnow 1974; Bishop et al. 1975), and in coal and uranium resource areas of Wyoming and other states in the Northern Great Plains (Montana Department of Mines 1979). Since the disturbed aquifers may be current or potential sources of water for municipalities, agriculture, or other users, the threat to water quality from these existing and potential activities has resulted in conflicts between water users and energy developers.

The second threat posed by energy developments to water quality is the disposal of large quantities of wastes. For example, near Canon City, Colorado, farming operations had to be curtailed in an area where excessive molybdenum seepage from uranium mill ponds contaminated nearby wells used for stock watering and irrigation (U.S., EPA, Off. of Drinking Water 1978, p. 67). Table 3-1 compares quantities of liquid and solid effluents produced by various energy conversion technologies

TABLE 3-1: Effluents From Energy Conversion Techniques

Technology	Size	Total Effluents[a]	
		(million short tons/year)	(pounds per 10^6 Btu)
Coal			
Power Generation	3,000 MWe		
Per Btu-electric[b]		0.53-2.55	15.6-79.0
Per Btu-thermal[b]			5.5-26.6
Gasification	250 MMcfd	0.45-1.96	10.9-47.8
Liquefaction	100,000 bbl/day	0.83-3.65	9.0-39.7
Oil Shale			
TOSCO II Oil Shale	50,000 bbl/day	16.19	365
Modified In Situ Oil Shale Processing	57,000 bbl/day	U	U
Uranium			
Underground Mine	1,000 mtpd (ore)	N	N
Surface Mine	1,100 mtpd (ore)	N-0.002	N-0.01
Mill	1,000 mtpy (yellowcake)	0.37	2.45
Solutional Mine-Mill	250 tpy	0.003-0.004	0.08

MWe = megawatt-electric
MMcfd = million cubic feet per day

U = Unknown
mtpd = metric tons per day

N = Negligible
mtpy = metric tons per year
tpy = tons per year

[a]Effluents include dissolved, wet, and dry solids; the range of values is that found at the six sites analyzed.

[b]To provide a basis for comparing power plants with synthetic fuel plants, effluents are expressed on both an electrical output basis and a per Btu of thermal input to the plant basis. See also the footnote to Table 2-7 and the associated discussion in the text.

on both a "standard-size-plant" basis and on a per-unit-of-energy-produced basis. Effluents from TOSCO II oil shale retorting represent by far the largest quantity, most of which is spent shale. A 50,000 barrel-per-day (bbl/day) TOSCO II surface retort produces about 16 million tons per year (MMtpy) of solids. Electric power generation compared on a British-thermal-unit (Btu) electric basis generates more effluents than any synfuel facility except TOSCO II and in situ oil shale. Of course, although the magnitudes of wastes are an important consideration, the chemical make-up of the wastes and the disposal method are also very important to water quality.

Table 3-2 summarizes some of the risks from land disruption and waste disposal and identifies control technologies for the six energy resources considered. Land disruption and waste disposal are treated in more detail below.

The CWA, Safe Drinking Water Act (1974), Resource Conservation and Recovery Act (RCRA 1976), and various other federal and state laws restrict the direct discharge of industrial pollutants into surface waters and groundwaters. Assuming these regulations are enforced, the most significant threat to water quality is from seepage, leaching, and runoff from waste disposal areas. For example, the current practice at most energy conversion facilities is to place dissolved and wet solid effluents in evaporative holding ponds, and possibly later to place the dried wastes in landfills. One risk is that berm failures or flooding of the holding pond can release large quantities of contaminants into local surface water. This recently occurred at a uranium mill near Gallup, New Mexico. In this instance, some 100 million gallons of liquid waste containing low-level radioactive material (uranium, thorium, and radium) and 1,100 tons of solid waste poured into a tributary of the Colorado River when a tailings pond dam broke July 16, 1979 (Kurtz 1979). In addition, seepage or leaching from the holding pond or landfill can contaminate groundwater. The degree of contamination will depend on the composition of the wastes, holding pond design, type of pond liner used (if any), pond management techniques, and the characteristics of nearby aquifers and of the soil overlying the aquifers. In turn, contaminated aquifers may introduce pollutants into local springs and streams.

The quality of water in a polluted surface stream will usually improve dramatically within one to two years after pollution sources are eliminated; however, polluted aquifers require 50 to 100 years or more to cleanse themselves, depending on geologic and soil conditions. Use of pond liners can significantly reduce the potential for leakage from ponds. Regulations to be

TABLE 3-2:   Summary of Major Water Quality Risks from
Land Disruption and Waste Disposal

Resource/ Extraction Technique	Selected Activites of Risk to Aquifers and Surface Waters	Control Technology
Coal		
Mining (surface and underground)	Dewatering of mines	Disposal of minewater in lined ponds
	Excavation or interception of aquifers	Adjusting groundwater movement
	Introduction of overburden, into aquifers	Sealing shafts
		Collection of runoff
	Acid mine drainage	
	Blasting	
	Disturbance of land surface	
Conversion	Waste disposal in ponds	Disposal of wastewater in lined ponds
	Disturbance of land surface	Collection of runoff in lined ponds
Oil Shale		
Mining	Dewatering of mines	Control of reclamation and drainage
	Excavation or interception of aquifers	Disposal of minewater in lined
	Disturbance of land surface	ponds
Modifed in situ	Mobilization of hazardous materials within aquifers	Sealing spent retorts
	Excavation of aquifers	
Surface retorting	Leaching of spent shale	Collection of runoff in lined ponds
	Disturbance of land surface	Sealing spent shale
		Stablization of spent shale
Uranium		
Mining	Exploration drill holes	Proper casing and cementing
	Disturbance of land surface	Control of runoff and drainage
	Excavation or interception of aquifers	Disposal of minewater in lined ponds
	Dewatering of mines	Collection of runoff in lined ponds
Solvent extraction	Release of hazardous substances	Control of solvent
		Injection of neutralizing agents
	Introduction of recovery agents	Flushing or pumping out of formation
Milling	Tailings pond leaching	Control of runoff and drainage
	Disturbance of land surface	Disposal of tailings in lined ponds
Oil, Gas, and Geothermal Drilling	Boreholes between aquifers	Proper casing and cementing
	Introduction of enchanced recovery agents (oil and gas)	Disposal of brines in lined ponds
	Blowouts	Installation of proper blowout preventors
	Brine reinjection or disposal	

Source: General Electric 1973; White et al. 1979a; and Grimshaw et al. 1978.

implemented by the Environmental Protection Agency (EPA) as a result of the RCRA (1976) will eventually place more stringent restrictions on the disposal of wastes from energy extraction and conversion facilities. Such waste generators may use more advanced methods of waste disposal such as incineration, solidification, or deep-well injection.

The activity of major concern in uranium, oil, natural gas, and geothermal energy exploration and development is probably the improper casing and cementing of boreholes. For example, improperly constructed oil wells, leaky casing, and improper cementing have resulted in hydrogen sulfide gas and poor quality water leaking into fresh water aquifers in the Bighorn Basin of Wyoming (Van der Leeden, Cerrillo, and Miller 1975). Similar activities have resulted in the flow of highly saline water from Crystal Geyser in Utah. Legislation has been passed in many western states describing procedures for abandoning and sealing wells including test holes, but there is a lack of adequate enforcement due primarily to the magnitude of the activity.

The degree of water pollution risks either from land disruption or waste disposal operations will vary greatly with the energy technology and the waste disposal technique. Because of both technology characteristics and the magnitude of future operations in the West, the greatest concerns in this study appear to be associated with coal, oil shale, and uranium development. The following sections briefly summarize specific water quality problems associated with these three resources.

Coal

Several activities associated with the extraction and conversion of coal pose risks to water quality in the West. Land disturbance, primarily from mining and erosion during construction and operation of conversion facilities, and waste disposal during coal mining and conversion are the major activities likely to degrade water quality.

Land Disturbance. For both the Upper MRB and the CRB extensive coal mining will at times pose a potential hazard to surface water and groundwater quality. A survey by Hounslow et al. (1978, p. 2) of eight surface coal mines located in New Mexico, Colorado, Wyoming, and Montana has determined that strip mining can increase levels of carbonates, sulfates, clays, and sulfides in runoff and leachate even in low rainfall areas by the increased movement of water through the mine's disturbed overburden, especially when the coal seam is associated with an aquifer. However, if coal is located above the aquifer, mining operations typically do not cause changes in groundwater quality unless significant precipitation filters through the soil (Hounslow et al. 1978, p. 179).

Increases in dissolved solids from surface coal mining depend on climate, groundwater flow, the composition/location of the aquifer and the disturbed overburden. In several mines in the Northern Great Plains, the groundwater has experienced an increase in total dissolved solids (TDS) content over premining conditions of 8,000 to 10,000 parts per million (ppm), making these waters generally unsuitable for domestic, agricultural, and most industrial uses (Montana Department of Mines 1979). These levels were caused by increased concentrations of magnesium, calcium, and sulfate. Thus far, such changes have not been detected in groundwater outside the mined areas. Nevertheless, as increased coal development occurs in the western U.S., some localities will experience area-wide disturbance to aquifers with salinity increases that may preclude some uses. Thus, enforcement of monitoring, control, and provisions for damages are provided in the Surface Mining Control and Reclamation Act (SMCRA 1977) regulations will be important to protect water quality.

Sediment loading of streams, resulting from erosion of exposed mine spoils or overburden, is a well recognized surface water quality problem. Sedimentation can have an adverse impact on aquatic ecosystems, and in some situations may change the morphology and stability of stream channels and flood plains so as to increase the frequency and severity of flooding. Sedimentation is a potential problem in all regions of the U.S. The Northern Great Plains and Rocky Mountain regions can experience severe sedimentation problems related to intense rainfall events, even though annual precipitation is low.

Salinity loading resulting from erosion of disturbed lands is also a difficult problem. A recent study by Rowe and McWhorter (1978) indicates that for coal surface mining in northwestern Colorado, annual salt loading from the disturbed land would be 2.13 to 2.37 tons per acre, a 500 percent increase above the premining rates. This value is still relatively small compared to annual salt loading from some irrigated agricultural lands which have been reported as high as 12 tons per acre in Colorado (Skogerboe and Walker 1972). The analysis also suggests that groundwater seepage from the disturbed areas accounts for more than 99 percent of the salt load from those lands. This is explained by the fact that the salinity concentration and the volume of groundwater seepage are both large relative to that for overland

runoff. Therefore, spoil management practices that promote overland runoff at the expense of percolation and
groundwater seepage will decrease the adverse salinity
impacts of mining. Other considerations, however, such
as maximizing soil moisture for revegetation, are not
necessarily compatible with a practice of enhancing overland runoff.

Waste Disposal. The magnitude and characteristics
of coal wastes will depend on the mining and conversion
processes and the coal and soil characteristics of the
site. Estimates of effluents which would be produced by
coal conversion facilities located near Gillette, Wyoming, are shown in Table 3-3. A Synthoil plant would
produce the most effluents, more than 2,500 tons per day
(tpd). The 3,000 MWe power plant will produce more than
2,300 tpd of solids, and Synthane and Lurgi plants are
expected to discharge more than 1,350 tons of solids per
day.
The solids dissolved in the waste stream are composed largely of chemicals from the coal mineral content
and cooling water. The principal dissolved solid constituents are calcium, magnesium, sodium, sulfate, and
chlorine; but, such wastes could also include smaller
amounts of many more harmful substances, including arse-
nic, boron, lead, mercury, selenium, and fluoride (White
et al. 1979b, vol. 2, pp. 843-45).
Wet suspended solids include flue gas scrubber
sludge, bottom ash, and cooling water treatment waste
sludge. Calcium carbonate and calcium sulfate are the
primary constitutents of flue gas sludge. Bottom ash
consists mainly of oxides of aluminum, silicon, and iron.
Calcium carbonate is the principal constituent of the
cooling water treatment sludge, which generally is small
in volume compared to the amount of bottom ash and flue
gas sludge. Dry solids produced by coal conversion processes are primarily fly ash, composed of oxides of aluminum, silicon, and iron.
Although many disposal configurations are possible,
some facility configurations include three types of disposal ponds and at least one type of landfill (Radian
1978a). Fly ash is dry and is usually deposited directly in a landfill. Bottom ash is usually sluiced to
an ash pond, allowed to settle, and the water recycled
or sent to an evaporation pond along with wastewater
from the demineralizer. Flue gas desulfurization (FGD)
sludge is usually routed along with cooling tower wastewater to a pond. Solids from the ash and sludge ponds
are periodically removed by dredging or other dewatering
techniques and deposited in a landfill. Fly ash, bottom
ash, and FGD sludge may be mixed together before land
filling to enhance compaction and stabilization.

TABLE 3-3: Effluents from Coal Conversion Facilities at Gillette

| Facility Type | Solids[a] (tons per day) | | | | Pounds Per Million Btu's in Product |
	Dissolved	Wet	Dry	Total	
Coal					
Lurgi (250 MMcfd)	25	1,186	154	1,365	12.1
Synthane (250 MMcfd)	25	321	1,014	1,360	12.1
Synthoil					
(100,000 bbl/day)	20	1,250	1,249	2,519	10.1
Electric Power[b]					
(3,000 MW)	50	958	1,303	2,311	
per Btu-electric					25.1
per Btu-thermal					8.8

MW = megawatt

Source: Radian Corporation 1978a; Gold et al. 1977.

[a]These values are given for a day when the facility is operating at full load. In order to obtain yearly values, these numbers must be multiplied by 365 days and by the average load factor. Load factors are typically 90 percent for synthetic fuels facilities and 70 percent for power plants. The values given as solids do not include the weight of the water in which the solids are suspended or dissolved.

[b]Concerning the Btu-electric and Btu-thermal measures, see the footnote in Table 3-1.

The environmental impact of disposing of such wastes is uncertain, but current information suggests that potential water quality problems do exist, especially when considering the magnitude of the wastes and the need for long-term containment of the more toxic wastes after a facility has shut down. For example, research sponsored by EPA (Geswein 1975) indicates that long-term capabilities of the various liner materials are unknown, in part because they can react with the wastes. Uncertainty is also due to inadequate information on long-term holding pond performance under different technological and locational operating conditions. In one case, a synthetic liner covered with one foot of soil was first used but was later found inadequate because heavy equipment could not enter the pond for cleaning. Soil cement was later used but was found to deteriorate severely. Asphaltic concrete was finally used satisfactorily to line five of six test ponds (Gavande, Holland, and Collins 1978, pp. 68-69). Another uncertainty about the environmental impact of coal conversion wastes is the toxicity of leachates and the rate of leaching under different conditions with a variety of wastes, types of liners and operation-management practices.

In 1974, the EPA began a field evaluation of the disposal of untreated and treated flue gas cleaning wastes near the Shawnee coal-fired power plant at Paducah, Kentucky (Leo and Rossoff 1976). In the clay lined ponds with low permeability, the groundwaters showed no evidence of altered quality. However, studies of the liquid beneath the pond but above the groundwater showed that the concentrations of TDS and major dissolved solids, (chlorides and sulfates) progressively increase in the leachate during the first year. The data also indicate that the concentrations may level off sometime during the second to fifth year. The concentrations of heavy metals in the leachate and the liquor show trends similar to those of chlorides and sulfates. However, it was not possible to project exact constituents and the concentrations that will occur in nearby groundwater because of the relatively small amounts present and the complex chemistry involved.

Failure of the holding pond berm is a potential source of pollution of local surface waters. The release of accumulated wet solids containing heavy metals, aromatic hydrocarbons, and various other potentially toxic chemicals could produce acute effects in exposed plants and animals. The design of holding pond berms must be site-specific, and failures are common in areas where previous design experience is limited or not available (see Smith 1973, p. 358). The quantities of wastes involved can be quite large. For example, at a typical site such as Gillette, Wyoming, the operation of a power plant and of Lurgi, Synthane, and Synthoil synthetic fuel facilities would produce more than 68 million tons of dissolved and suspended waste matter over a twenty-five-year period.

Seepage from holding ponds can also contaminate groundwater. The degree of contamination depends on the composition of materials in the ponds, the pond and liner design, pond-management techniques, and the characteristics of nearby aquifers and of the soil overlying the aquifers. Once contaminated, aquifers may introduce pollutants into local springs, seeps, and streams. The problem may be a long-term one. Although water quality in a surface stream will usually improve dramatically within one to two years after the sources of pollution are eliminated, aquifers require much longer periods to cleanse themselves, depending on geologic and soil contions.

Oil Shale

Oil shale extraction can be accomplished by mining the shale and then retorting it in facilities on the surface or by in situ (or modified in situ) methods. Although both groundwater and surface waters can be adversely affected, the impacts on groundwater systems are expected to be greater.

Land Disturbance. The extent of land disturbance varies with the type, size, and duration of the mining and retorting operations. As with most mining operations, the construction phase activities could impact local surface waters by increasing sediment load. The severity of the surface water impact from erosion is dependent on the size of the area impacted, soil characteristics, drainage characteristics of the basin, and precipitation levels. To illustrate the magnitude of mining operations, the production of 400,000 bbl/day of shale oil from surface retorting will require mining approximately 250 million tons of oil shale annually. After thirty years of operation this will mean that roughly 40,000 to 50,000 acres (or 60 to 75 square miles) will be affected by the surface and subsurface activities of underground mining.[1]
Water quality could be affected in three ways. First, this level of mining could result in mine dewatering discharges of from 3,500 to 30,000 acre-feet per year (AFY) depending on the location of the mines. Depending on treatment and disposal of mine water, surface water quality could be affected since the mine water may have a concentration of up to 40,000 milligrams per liter (mg/l) TDS and high levels of some trace elements. For example, fluoride levels of 10 mg/l to 20 mg/l have been measured, which exceed drinking water standards by a factor of ten (Crawford et al. 1977, p. 121). Options for disposing of mine water include reinjecting the water or spray irrigating the excess water on the shale spoil pile. A second potential problem is that groundwater from lower aquifers with salinities up to 40,000 mg/l could intrude into the fresher, dewatered aquifers. This may reduce the quality and use of the fresher groundwater and surface waters of streams such as Piceance Creek and Parachute Creek. A third but highly uncertain problem in mined formations is the release of trace elements which become more soluble when air is introduced, subjecting minerals to oxidation and leaching.
With in situ extraction, the disposal of minewater and the intrusion of saline groundwater into fresher aquifers are also potential problems. In addition, during the operational phase, shale oil and related organic and inorganic compounds could all be considered

potential contaminants that could enter the groundwater
system. After the operations cease, the in situ retorts
can become saturated by groundwater, releasing contaminants into the water and lowering water quality. The
higher permeability of the retort area can also be expected to affect the flow relationship among aquifers.
The combustion of the oil shale in these retorts may
mobilize trace elements, organic substances, or other
potentially hazardous materials that could be subsequently leached by groundwater after operations cease
and the water table reestablishes itself. Whether the
high temperatures used in the retort will increase or
decrease the availability of trace elements or other
contaminants is unknown.

Waste Disposal. Disposing of an enormous volume of
spent shale from the surface retorting process could
introduce chemicals into surface and groundwater. Over
a thirty-year period, the production of 400,000 bbl/day
from surface retorting would result in enough spent
shale to cover roughly 24 square miles to a depth of 100
feet. Spent shale can pose more serious water quality
problems than raw shale. Salt mobilization is roughly
four times greater in spent than in unprocessed shale
bodies, resulting in a potential long-term source of
saline surface and groundwater. The spent shale from
the TOSCO II process also contains one to two percent
organic matter with identified but unknown proportions
of carcinogenic hydrocarbons (Schmidt-Collerus 1974).
    Because of the large volumes involved, energy com-
panies using surface retorts have only two options:
store all the spent shale above ground or return part of
the waste to the mine and store the remainder above
ground. Generally, surface retorting projects have proposed to use above-ground storage (Crawford et al. 1977,
pp. 24-55). The spent shale will be accumulated in can-
yons behind strategically placed dams. This would force
runoff water to flow through drainage channels into
catchment basins and allow drainage from surrounding
land to bypass the storage pile.
    Spent shale apparently will not be classified as a
hazardous waste. A leaching test of Paraho processed
shale by Denver Research Institue (1979, sec. 7, p. 5)
showed that all toxic elements extracted were less than
the allowable standards. Further, since the residual
carbon in the spent shale has been converted to coke in
the retorting process, the organic matter it contains
should not be readily leached out by rainfall. If this
projection should be incorrect, it is possible to remove
the carbonaceous residue by further treatment of the
spent shale.

Studies of spent shale disposal have shown that shale piles can be revegetated either by artificial leaching with 7 3/4 inches of applied water or by three years of natural snow and rainfall (Cook 1979). Covering with topsoil is helpful but not essential for revegetation. However, the productivity of the revegetative stage will be higher with adequate restoration of top soil.

Studies of runoff have shown that the maximum salinity in initial observations was two and one-half times that of normal surface runoff (Harbert and Berg 1978 p.4). Continued observations on these experimental plots have shown decreasing shale runoff salinity with time (Cook 1979) as might be expected from both weathering of the spent shale and decreasing concentrations of salts.

Uranium

In the Upper MRB and the CRB the production of uranium could affect both surface water and groundwater. The severity of the problems associated with waste disposal and land disturbance vary according to the mining process. The two alternative uranium mining methods considered here are in situ solution mining and underground mining with surface milling.

Land Disturbance. The severity of land disturbance by uranium mining depends on the technique used to remove the ore. In situ solution mining is characterized by a lack of major surface disturbances and the absence of large quantities of mine wastes.[2] However, as with all mining activities, land clearing and grading during the construction and operating stages increase the potential for erosion. Sediment introduction into nearby streams can be expected to be small with the in situ method of mining because the total acreage disturbed is relatively small. With conventional uranium mining and surface milling, the greatest threat occurs during the operational phase from ore extraction and milling and from disposal of wastes. Surface runoff from overburden and waste dumps may contain appreciable amounts of dissolved solids not contained in normal runoff for the area. Runoff of this kind could significantly impact local water quality.

Land disturbance can also lead to an altering of groundwater quality. In situ solution mining poses a greater risk for groundwater quality than does conventional mining. During the mining process the degradation of groundwater adjacent to the ore zone is of the greatest concern. In the mining process the leachate solution is pumped underground to a uranium ore body.

The solution is typically alkaline, such as ammonium bicarbonate. The loaded leachate solution is then pumped to the surface where the dissolved uranium is removed. Pumping rates are controlled to prevent the flow of the solution from the outer injection wells to the inner wells. However, accidental intrusion of the solution into groundwater zones adjacent to the ore zone during the mining process would at least temporarily degrade groundwater quality until the solution could be recovered or diluted with uncontaminated water.

Waste Disposal. Waste disposal problems will vary according to the process being used. Liquid effluents and solid wastes, both radioactive and nonradioactive, are generated during the solution mining process. The wastes produced are not as large as those produced by conventional mining methods. The largest volumes of solid waste produced in the solution mining process are calcium carbonate or calcium sulfate, depending on whether alkaline or acid leach methods are used. At one mining site in Wyoming, it is estimated that two pounds of calcium carbonate are produced for every one pound of uranium oxide. The solid wastes are in solution and are temporarily stored in lined evaporation ponds. Radioactive solid wastes must eventually be disposed of in a licensed uranium mill tailings disposal site (Stone & Webster 1979a, pp. 3-20, 3-21).

Most uranium in situ mining activities will be designed for zero level discharge of contaminated wastes to surface waters. Any surface discharges must be in accordance with the National Pollutant Discharge Elimination System (NPDES) guidelines of the CWA. However, effluents could be released accidentally through flooding, failure of evaporation pond dikes, or seepage through inadequate pond liners. In cases where underground aquifers are naturally connected to surface waters, improperly abandoned uranium wells and test holes can provide a mechanism whereby contaminants reach surface waters.

Liquid effluents and solid wastes will also be produced in conventional mining and milling operations. In most cases, liquid wastes from the mill are mixed with spent tailings to form a slurry and then transported to the tailings disposal area. As with uranium in situ solution mining, most operations will be designed for zero level discharge of pollutants into surface waters. However, water quality impacts could result from the seepage of wastes from tailings ponds. The amount of seepage depends on the size and depth of the tailings pond, construction characteristics, soil characteristics, and liner quality. For example, during the period of 1960 to 1973, waste disposal practices at two mills in

the Grants Mineral Belt (New Mexico) introduced an estimated 200,000 kilograms of dissolved uranium into the
subsurface waters through seepage and direct injection
(Kaufmann, Eadie, and Russell 1977). Water samples in
locations in the Grants Mineral Belt show elevated
levels of radium-226, although it is difficult to determine if this is the result of natural occurrence or uranium production activities.

As shown in Table 3-4, the largest amount of liquid
effluent produced at a typical mining and milling operation is from mine dewatering. Mine dewatering is necessary when the mine penetrates the local groundwater table. This procedure for removing water which intrudes
into the mining area can affect surface and groundwater
by (1) lowering the local groundwater table; (2) altering water quality, especially if the mine water is of a
lesser quality than the local surface water supply; and
(3) increasing local surface water flow (Radian 1978b).
The magnitude or severity of these changes depends on
the scale of operation and total volume of mine dewatering necessary.

Much uncertainty exists regarding the potential ra-
diological dangers of abandoned tailings ponds and bore
holes (see box "Uranium Tailings Concerns in the Upper
Colorado River"). In some instances, only 10 to 15 percent of the radioactive material is removed from the
ore during the milling process, the remainder staying in
the tailings primarily due to unextracted radium. The
initial design and long-term maintenance of tailings
ponds are critical elements in assessing the water
quality risks of a mining and milling site.

Uranium Tailings Concerns in the
Upper Colorado River

An example of the uncertainty involved in the
radioactive seepage debate is encountered in the Upper
CRB. "Uranium mill tailings piled throughout the Upper
CRB are seeping into streams and groundwater, but state
officials are divided on the severity of the problem....
Some amounts of radioactivity from the abandoned Vitro
Uranium Mill tailings are probably seeping into the
Jordan River near Salt Lake City.... How much and what
effect it is having on the area's water supply are not
known...." The Vitro site in Salt Lake is one of more
than a dozen abandoned uranium mill tailings sites in
the Upper CRB.

-- High Country News 1980, p. 13.

TABLE 3-4:   Liquid Effluents from Uranium Mining
             and Milling

Source	Discharge (AFY)
Excess mine dewatering	
1,200 tpd open pit mine	500-1,500
1,200 tpd underground mine	4,839
Mill process water (tailings solution)	600[a]
Total	5,939-6,939

Source: White et al. 1979a, vol. 4, pp. 4-7.

[a]Evaporated from tailings pond.

Water Quality Policies

A variety of federal and state laws protect the nation's surface and groundwater resources. For energy
development, controls can be placed into two categories:
those controlling the discharge of pollutants into surface or groundwater and the disposal of wastes; and
those controlling the disruption and/or contamination of
aquifers from mining and other in situ recovery processes. The most important laws and regulatory programs
are discussed below.

Pollution Discharges and Waste Disposal. At the
federal level, effluent discharges into interstate or
navigable waters from industrial sources, including
energy facilities, are primarily regulated through the
CWA (1977) which includes provisions of the Federal
Water Pollution Control Act (FWPCA) of 1972. The program
to regulate point sources under the CWA is known as the
NPDES.   Under this program, no effluent can be discharged by a point source without a permit which sets
the conditions under which the discharge may be made.
Table 3-5 summarizes the regulations affecting energy
facility discharges.   Permits are issued by EPA or the
state if the state program has been approved by EPA.   To
be approved by EPA, a state's laws must authorize an
agency to:

TABLE 3-5:   Federal Effluent Regulations[a] Affecting
             Energy Facilities

Facility or Pollutant	Treatment Required
Industrial	"Best practicable control technology cur-rently available" as defined by the EPA administrator by July 1, 1978.[b]
	"Best available technology economically achievable" as determined by the EPA administrator by July 1, 1984.[b] Limits are based on categories or classes of industries.
	National performance standards--including zero-discharge, if practicable--for each new category of source.
Toxic pollutants (seriously harmful to human or other life)	Effluent limitations including prohibition of discharge, if needed, to provide "an ample margin of safety" set by the EPA administrator[c].
Thermal discharge	Effluent limitations to ensure a balanced population of fish, shellfish, and wild-life.
Oil or hazardous substances	No discharge into U.S. waters, adjoining shorelines or contiguous zone waters.

[a]FWPCA 1972. Adapted from Congressional Quarterly (1973, p. 797).

[b]Data here reflect the change in the law made by the CWA (1977).

[c]EPA was directed by court action to issue water quality criteria for sixty-five organic and inorganic chemicals (NRDC v. Train 1976). About twenty of these chemicals are associated with energy extraction and conversion technologies (Chartock et al. 1981).

(1) Issue permits that apply all CWA requirements;

(2) Monitor permittees to the extent required by the CWA;

(3) Notify the public when an application is made and provide an opportunity for a public hearing;

(4) Report to EPA the downstream states that would be affected, and present the U.S., Army Corps of Engineers an opportunity to object; and

(5) Impose requirements on publicly owned treatment facilities

Four of the eight states in our study area have EPA approved programs: Colorado, Montana, North Dakota, and Wyoming. However, even when EPA has approved a state program, it still can veto any individual permit proposal and withdraw its approval of a state's entire permit program.

Permit holders under NPDES must meet effluent standards for specific industries or more stringent ambient water quality limitations (CWA 1977, §303). Streams, or segments of streams, can be designated as being either "effluent limited" or "water quality limited" (U.S., EPA 1974, p. 8); discharges into a "water quality limited" stream are subject to more stringent limitations.

Despite these discharge controls, regulations to restrict discharges into groundwater may be inadequate. In some cases these discharge controls have not provided comprehensive jurisdiction because of legal challenges. For example, six uranium companies in New Mexico challenged the EPA's authority under the CWA primarily on the basis that temporary streams receiving mine water discharges are not navigable waters and that the effluent limitations are arbitrary.

Existing point source control requirements affect energy development technologies by requiring that discharges be cleaned up or retained in holding ponds. The costs of supplying water and cleaning up discharges to meet the effluent standards may make it more economical for developers to continue to treat and recycle the water as long as possible and retain pollutants in evaporation ponds rather than discharge treated effluents to surface waters. However, as described previously, this approach does not necessarily solve the water quality problem because these wastes accumulate in the ponds and can create potentially significant surface water and groundwater quality problems. Thus, the CWA requirements aimed at protecting surface water quality contribute to the decision to use alternative means of effluent disposal that can, in turn, lead to other potentially serious water quality problems.

Until recently, the disposal of such wastes was largely unregulated. However, the 1976 RCRA now places holding ponds and land fills under federal control. RCRA addresses needs in the following principal areas: (1) development of a system for the management of solid waste; (2) development of a system for the management of hazardous waste and; (3) recovery of resources from solid and hazardous waste.

Regulations under RCRA dealing with hazardous wastes and state and regional solid-waste plans are still being developed. It is unknown whether some of the solid wastes generated by energy facilities will be classified as hazardous. The definition of "hazardous waste" contained in Section 1004 (5) reads:

> The term 'hazardous waste' means solid waste or combinations of solid waste, which because of its quantity, concentration or physical, chemical, or infectious characteristics may: (a) cause, or significantly contribute to an increase in mortality or an increase in serious, irreversible, or incapacitating, reversible, illness; or (b) pose substantial present or potential hazard to human health or the environment when improperly treated, stored, transported, or disposed of, or otherwise managed.

Regulations proposed by EPA include classifying fly ash, bottom ash, and flue gas scrubber sludge in a category called "special wastes." For special wastes, EPA has proposed postponing the promulgation of standards until June 1982 and their implementation until June 1983. If these special wastes are eventually subject to strict hazardous waste controls, the costs of hazardous waste disposal by energy industries could increase from the present rate of about $2 a ton to as much as $90 a ton (EPRI Journal 1978). These estimated cost increases are due to requirements such as siting hazardous waste disposal facilities away from wetlands, 100-year flood areas, critical habitats, active fault zones, public roads, and residences. A disposal facility located over usable aquifers may be required to have at least ten feet of low permeability soil between it and the aquifer and have groundwater and leachate monitoring systems.

The Safe Drinking Water Act (1974) protects groundwater quality and surface water quality from any actions which pose a risk to drinking water supplies. Two programs under the Safe Drinking Water Act may place limitations on pollution resulting from energy development: (1) the underground injection control (UIC) program and; (2) the surface impoundment assessment (SIA) evaluation system. The UIC program is designed to protect groundwater quality from the injection of unwanted brines brought to the surface during oil, gas, and geothermal energy production (U.S., EPA, Office of Drinking Water 1979, p. 1). Final regulations were issued early in 1980 (40 C.F.R. 122 and 146). Surface impoundments including industrial waste ponds were being surveyed under the SIA evaluation system to determine their threat to groundwater quality. A regulatory strategy to protect groundwater quality from such impoundments is being developed (Josephson 1980, pp. 38-44). Regulations under the Safe Drinking Water Act have not yet been widely applied to energy production operations, however, and the act explicitly states that regulations will not be established that would interfere with the production of oil or natural gas unless such regulations are deemed essential to prevent endangering underground sources of drinking water (Shaw 1976, p. 532).

EPA is authorized through the Toxic Substances Control Act (1976) to regulate the disposal of toxic materials by manufacturers, users, transporters, and commercial disposal firms. This act is primarily directed toward the control of the chemical manufacturing industry, but does provide controls related to energy processes, including synthetic fuel plants. The Toxic Substances Control Act provides for a federal program and is administered primarily through EPA's Office of Toxic Substances.

Water Quality Protection from Land Disruption. Regulations established to control the environmental effects of surface coal mining under the 1977 SMCRA include protection of groundwater resources. However, there is still considerable uncertainty as to how effective the controls will be and how rigidly the regulations will be enforced. In addition, there are no similar comprehensive regulations for ensuring water quality protection for other resources, such as uranium and oil shale mining and geothermal energy production.

The SMCRA has several sections that specifically protect surface water and groundwater quality affected during coal mining. For example, exploratory drill holes must be cased to prevent surface water and groundwater contamination (30 C.F.R. 816.13-15); and if needed, surface water and shallow groundwater may be diverted around mined areas to minimize the need for water treatment technology within the mine (30 C.F.R. 816.43).

In general, certain steps must be taken to protect the hydrologic balance and quality of groundwater (30 C.F.R. 816.50-51). These provisions include:

(1) The proper placement of backfilled materials to minimize contamination of groundwater;

(2) The appropriate location and design of cuts and excavations to preserve groundwater quality for post-mining use; and

(3) The restoration of the aquifer to approximate pre-mining recharge capacity.

Special protection for alluvial valley floors is also required. This provision is primarily applicable to the arid and semiarid areas of the country (30 C.F.R. 822). To obtain a permit for mining, companies have the burden of proof to demonstrate that "...operations shall not cause material damage to the quality or quantity of water in surface or underground water systems that supply alluvial valley floors...."

North Dakota serves as one example of state implementation of the federal surface-mining reclamation act. In April 1979 the governor signed into law the state Surface Mine Reclamation Act (N. Dak. Century Code, Ch. 38-14.1). For example, it provides procedures to declare lands unsuitable for mining if operations would result in a substantial loss or reduction in productivity of the long-range water supply, including aquifer and aquifer-recharge areas (N. Dak. Century Code, Ch. 38-14.1-05). When a mining permit is obtained, the applicant must indicate the probable hydrologic consequences of the mining and reclamation operations both on and off the mine site, including changes in quality of groundwater under seasonal flow conditions (N. Dak. Century Code, Ch. 38-14.1-14.1.o). A description of protective measures is also required, including processes to assure protection of the quality of groundwater (N. Dak. Century Code, Ch. 38-14.1-14.2.i). A bond is required to obtain a mining permit, and this bond can be released only after the mining commission determines the extent of groundwater pollution and, if needed, the cost of abating the pollution (N. Dak. Century Code, Ch. 38-14.1-17). Alluvial valley floors are protected, and the Mining Corporation can delete certain areas from a strip mine permit if it finds that water pollution may result. Specific reclamation procedures are required if groundwater is threatened (N. Dak. Century Code, Ch. 38-14.1).

States have jurisdiction for protecting groundwater in areas of oil and gas operations and of uranium development. However, uranium mining operations occur largely on patented or private lands, and control by state agencies is limited. Statutes in most states do not provide the authority to control uranium mine dewatering or subsurface waste injection. For example, the Water Quality Control Division in the Colorado Department of Health (1979) has recommended that monitoring programs be implemented in hardrock mining, including uranium development. The Uranium Mill Tailings Radiation Control Act of 1978 was passed to provide for the environmentally safe disposal of uranium mill tailings including remedial action at existing sites and long-term disposal safety after termination of operations. Cooperative agreements by the Secretary of Energy with the states and Indian tribes are authorized to acquire land, provide financial assistance, and undertake remedial action to safely dispose of uranium mill tailings. However, this Act along with the Atomic Energy Act of 1954 does not provide for safeguards to surface water and groundwater during uranium mining and milling to the extent that the SMCRA does for coal. This is partly because regulatory action depends to a large extent on states or Indian tribes. For oil and gas development, control is typically vested in oil and gas commissions or in a corporation commission that is responsible for "efficient production." State laws generally call for the use of state-of-the-art cementing, casing, and plugging techniques to minimize the possiblity of aquifer contamination. For water supply wells used in conjunction with energy facilities, states typically have provisions protecting water quality by licensing and regulating well construction and abandonment (Van der Leeden 1973, p. 18).

In summary, not only is there scientific uncertainty as to the long-term water quality impacts of energy development in the West, but there are also gaps and uncertainties in the regulatory programs. As stated by one state water quality agency, one of the greatest needs in the area of water quality management for active mining operations is for regulatory coordination and program consistency among the several states and federal agencies (Colo. Dept. of Health, WQCD 1979, p. 24).

Area-Wide Water Quality Management Planning. Section 208 of the CWA establishes procedures under which states or regional agencies are required to establish long-term water quality plans and "nonpoint source" regulatory programs. Irrigated and nonirrigated agriculture, mining, urban runoff, and rural sanitation are examples of nonpoint sources. However, the overall effectiveness of the 208 program has been criticized, and application to energy projects in the West has been especially limited.

Section 208 also requires governors to designate areas that have significant water quality problems. Responsibility for planning is delegated to local governments, coordinated by the designated 208 area-wide planning agency (a regional council of governments, for example). Completed plans must be approved by the state and EPA prior to implementation.

Of special relevance to energy development are 208 projects involving groundwater and mining. EPA funded twenty intensive groundwater protection projects, ten each in fiscal year (FY) 80 and FY81. The Larimer-Weld Regional Council of Governments, for example, received $1.3 million under the 208 prototype planning function in order to develop a comprehensive groundwater management program (the Denver Basin Project). Overall funding for 208 planning after FY80 is allocated only for nonpoint source control programs. EPA will report to Congress in 1983 regarding the federal role in nonpoint source control program implementation (U.S., EPA 1979, 1980; Costle 1979).

SALINITY CONTROL

Summary: Salinity will continue to be an important issue in the CRB and in the Yellowstone River Basin. Increases in salinity from energy development are expected to be small relative to the contribution of natural salt flows and runoff from irrigated agriculture. Nevertheless, energy development will intensify conflicts over salinity control, and energy production could be constrained by salinity standards if adequate controls are not established. States of the CRB have adopted salinity standards for the Colorado River designed to achieve no salt return, wherever practicable, from industrial discharges. However, it is unclear what the states can or will do to enforce these standards. The Environmental Defense Fund (EDF) has filed suit against the EPA claiming failure to comply with the CWA's salinity control requirements.

Increasing levels of salinity in both the CRB and the Yellowstone River Basin are important for both environmental and economic reasons. Each mg/l of increased salinity at Imperial Dam has a cost to the environment, to municipal water users, and to agricultural production of about $230,000 in 1972 dollars (U.S., DOI, BuRec and Dept. of Agric., SCS 1974). This figure is being updated and a tentative salinity cost of $343,000 per mg/l increase is being used by DOI (Kauffman 1980).

Along the Colorado River, salt concentrations and loadings generally increase downstream because of diffuse salt pickup from undisturbed lands, agricultural runoff, and saline springs. Figure 3-1 shows 1976 salinity concentrations at a few points along major rivers of the CRB (U.S., DOI, BuRec 1979). In the Upper MRB, the subbasin of most concern with regard to salinity is the Yellowstone. Average salinity concentrations for various locations in the Upper MRB are shown in Figure 3-2 (Klarich and Thomas 1977). Salinity problems are most severe in the Powder River.

Economic development of the CRB has been the main cause of the increased salinity--primarily two and one-half million acres of agricultural irrigation and industrial and municipal uses. In 1970, about 69 percent of the total salt load came from natural sources, about 30 percent came from agriculture, and about one percent came from municipal and industrial sources. (Utah State U., Utah Water Research Lab. 1975, pt. 1, p. 66).

In 1975 the states of the CRB agreed to limit salinity concentrations at three points to 1972 levels. The 1972 "numeric criteria" and the annual average salinity at these stations are shown in Table 3-6. A consistent decrease in salinity has recently occurred (CRB Salinity Control Forum 1978a, pp. 7-8, 24). Apparently the major cause of the decrease is construction of the Colorado River Storage Project reservoirs. Lake Powell seems to be of particular importance since that reservoir was completed at about the same time the relationshp between salt concentration and flow changed. One explanation is that salt loading may have been reduced by the construction of these reservoirs in the following ways (CRB Salinity Control Forum 1978a, pp. 21-22):

(1) Shorter stream channels and inundation of smaller areas have lessened salt accumulation from overflowing water;

(2) Widely fluctuating flows are controlled by the reservoirs which have greatly reduced floods. This reduces erosion and salt pickup from these areas;

(3) Suspended solids in high velocity rivers are now largely retained in reservoirs in bottom sediments. These sediments no longer release salts to the rivers; and

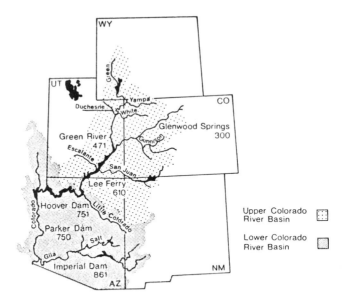

Figure 3-1: Salinity in the Colorado River Basin
(mg/l)

Figure 3-2: Salinity in the Upper Missouri River Basin
(mg/l)

TABLE 3-6: Flow Weighted Annual Average Salinity[a] (milligrams per liter)[b]

Station	1972 Salinity Levels (Numeric Criteria)[b]	1973	1974	1975	1976	1977
Hoover Dam	723	706	686	689	675	667
Parker Dam	747	726	700	703	699	689
Imperial Dam	879	846	836	829	823	820

[a]CRB Salinity Control Forum 1978a, p. 7.

[b]These standards were subsequently approved by EPA.

(4) Dissolved salts are apparently being chemically precipitated in the reservoirs, especially in Lake Powell.

A second explanation for decreased salinity concentrations is the salt cleanup projects in the basin. For example, New Mexico authorities have suggested that the Mohawk Clean-Up Project in Arizona may be responsible for some of the salinity decrease (Gilliland 1979). It has also been suggested that salinity concentrations in the river have decreased because of flow variations in recent years.

This short-term trend of decreasing salinity has led some, including the states of the Upper CRB, to conclude that there is no salinity problem at the present time. However, at least one projection of the effects of increased water use in the CRB indicates that salinity concentration levels will increase in the future. If this is true, new salinity control efforts will be necessary before the year 2000 to keep salinity below 1972 levels.

Effects of Energy Development

Energy development can contribute to salinity increases in two ways:

- Salt loading due to runoff from surface disturbances (such as oil shale disposal,
  mining, and roads) and disturbances to
  aquifers which feed surface streams; and

- The concentrating effects of water consumption
  by energy conversion facilities.

Considerable uncertainty exists about the salt loading
effects of energy development. The amount of salt loading will depend on soil type, precipitation levels, degree of reclamation success, and degree of runoff diversions and controls from mined areas. As discussed above,
surface coal mining runoff from reclaimed areas is diverted into catchment basins and then either treated or
placed in holding ponds to minimize salt additions to
surface water. With oil shale, the initial level of
salinity in runoff from spent shale piles is estimated
to be about 2.5 times that for normal surface runoff.
These levels will decrease with time and most shale development plans call for controlling runoff from disposal areas. However, there is also concern that dewatering aquifers during oil shale mining might allow saline
groundwater in deeper aquifers to intrude into dewatered
fresh aquifers. Some of this saline water could then
flow into surface streams. Despite such uncertainties,
it is generally believed that salinity increases due to
salt loading in surface waters will be small compared to
the concentrating effects of consumptive water use.
    Several models have estimated the salinity concen-
tration increases for the CRB from consumptive water use
by energy facilities. The results must be viewed with
caution because they are based on assumptions which may
not be accurate. Thus, current salinity models may
overestimate or underestimate salinity effects by as
much as 100 percent. Salinity concentrations depend on
flow levels, which vary year to year, and the rate and
scale of development. Nevertheless, such models indicate
the magnitude of effects from either a single energy
facility or from energy development on a regional basis.
    For example, one study estimated the effect of two
coal-fired power plant units, totaling 760 megawatts,
planned near Craig, Colorado. These two units, consuming 12,000 AFY, were estimated to result in an average salinity increase of 0.7 mg/l below Hoover Dam
(U.S., DOI, BLM 1976, pp. III-3 through III-9).
Another study estimated that a 410 MMcfd coal gasification development in the Four Corners region of New
Mexico, consuming 15,000 AFY from the San Juan River,
would increase salinity about 2.4 mg/l at Imperial Dam
(U.S., DOI, BuRec 1977).

One model of water use for energy development in the twenty-one subbasins of the Upper CRB determined that water consumption from high salinity streams could act- ually improve downstream salinity, This study iden- tified the Gunnison, Dirty Devil, San Rafael, Price, Duchesne, and Black Fork rivers as optimum locations for water supplies for consumptive use. If water resources for energy development in the Upper CRB were taken from these streams, a net improvement of about 20 mg/l would occur at Lees Ferry (Flug et al. 1977, p. VI-3).

Another comprehensive study on salinity effects for the CRB used three levels of energy development (low, medium, and high) to determine the amount of water that would be used by energy, agriculture, and for export (Utah State U., Utah Water Research Lab. 1975, pts. 1 and 2). Based on 1972 conditions, projections were made for 1977, 1983-1985, and 1990-2000 for three flow levels. Table 3-7 summarizes the results of this study, assuming a virgin flow of about 14 million acre-feet per year (MMAFY) at Lees Ferry and a moderate amount of water exported from the Basin.

Case A in Table 3-7 shows the salinity effects of a medium amount of water usage for both agriculture and energy development. If case A is the "base" condition, then case B--with medium agriculture water use and high energy water use--shows the effects of energy develop- ment on salinity. The salt load in case B decreases relative to the base case, because of water (and there- fore salt) withdrawals. In case B the flow at Lees Ferry, Arizona, is projected to decline by about 1.73 MMAFY in the 1990 to 2000 period, due to increased withdrawals for energy development. Therefore, even though salt loadings decline, salinity concentrations increase because of the reduced flow.

Comparing case A with cases C and D shows what might happen if water is shifted from agriculture to energy or if agricultural water use is reduced. For cases A and C, salinity concentrations are projected to increase, but salt loading is less for case C. Case C includes more use for energy and less for agriculture. However, in case D the salt load on the river basin is decreased even more than case C compared to case A where increased energy consumption of water is moderate and agricultural usage of water is low.

A projection of future salinity levels has also been made for the Upper MRB (Klarich and Thomas 1977). This study used three levels of development (low, intermedi- ate, and high) for energy, irrigation, and municipal wa- ter use by subregions of the Yellowstone River and pre- dicted the monthly effects on salinity concentrations of reduced stream flows. The conclusions of this analysis are that in the eastern portion of the Yellowstone River Basin, energy development and the resulting reduction in

TABLE 3-7: Predicted Salinity Effects at Imperial
           Dam for Alternative Energy Development
           Futures[a]

Case	Water Use[b]		Salinity					
			1977		1983-1985		1990-2000	
	Agriculture	Energy	Mtpy	mg/l	Mtpy	mg/l	Mtpy	mg/l
A (base)	M	M	920	770	918	895	780	1,045
B	M	H	920	780	915	940	760	1,140
C	L	H	920	905	910	905	750	1,055
D	L	M	920	875	913	875	775	1,000

Mtpy = thousand tons per year

[a]Adapted from Bishop 1977, p. 667.

[b]Resource utilization level: H = high, M = medium (most likely), L = low. The
development levels projected in this study and the amounts of water used by
agriculture, energy, and other sectors were not given. Instead, flow at Lees
Ferry (MMAFY) under two development levels were presented as follows:

		Year	
Case	1977	1983	1990
A	10.471	9.924	9.177
B	10.468	9.821	8.740

stream flow can have a major adverse effect on surface
water quality. Increased salinity is predicted to be
most severe in the Tongue and Powder River subbasins
under any level of development; in the Lower Yellowstone
subbasin in Montana (from Miles City to Sidney), it will
be severe if a high level of development occurs. In the
western portion of the Yellowstone Basin the effects of
energy development on salinity are not projected to be
severe. In the Upper and Mid-Yellowstone subbasins and
the Bighorn subbasin the only projected salinity problems will occur during late summer and early fall if
there is a high level of water use or during dry years.

The Salinity Control Policy System

The most serious surface water salinity problems in
our eight-state area are in the CRB. Consequently, the
regulatory structure in this basin is the most developed
and the basin's political/social system is the most sensitive to the issue of salinity control. Salinity levels

in surface waters have not been much of a public issue in the Upper MRB until recently. The State of Montana has instream standards of 500 mg/l for salinity, but these are currently undergoing revision for the Tongue and Powder rivers (Boris and Krutilla 1980, p. 241).

Section 103 of the 1972 FWPCA (now the 1977 CWA) requires that the states of the Colorado River (and others) establish a mechanism for interstate cooperation in setting numerical criteria for salinity control. In response the CRB Salinity Control Forum was formed in November 1973 by the seven basin states. Shortly thereafter these states agreed not to allow salinity in the river at Imperial Dam to increase above the 1972 level. This standard was subsequently approved by EPA.

Also in 1973, the U.S. entered into an agreement with Mexico, the effect of which was to limit the salinity of water from the Colorado River flowing into Mexico. The agreement requires that the salinity level of water delivered to Mexico shall be no greater than 115 ppm[3] (plus or minus 30 ppm) above the annual average salinity of water entering Imperial Dam (International Boundary and Water Commission 1973). This means that currently the maximum salinity concentration that can be delivered to Mexico is about 970 ppm.

The CRB Salinity Control Act of 1974 authorized the development of a major salinity control project near Yuma, Arizona. This project includes a desalination plant and other activites such as lining irrigation canals and reducing irrigated acreages. This project is being handled by the Water and Power Resources Service (WPRS), formerly the Bureau of Reclamation, and is designed mainly to help the U.S. comply with the agreement with Mexico. However, actions to implement the project have been delayed and estimated costs have risen. For example, the estimated cost of the Yuma desalination plant alone increased by 187 percent (from $62,080,000 to $178,400,000) between 1974 and 1977, despite a decrease in the design capacity of the plant to the minimum size allowable by the Act. Overall estimated costs for the Yuma project increased by 115 percent (from $155,500,000 to $333,692,000) during this same period. Costs per acre-foot are expected to average $338 for 88,000 AFY processed. Recent estimates are that the plant will cost $223 million (January 1980 dollars) and the total project will cost $369,560,000 (October 1979 dollars) (Kauffman 1980). And even if the plant is built, WPRS estimates an additional 35,000 AF of water must be replaced in order to satisfy flow and salinity requirements of agreements with Mexico (U.S., GAO, Comptroller General 1979, pp. 38-40).

The CRB Salinity Control Act (1974) also authorized four other salinity-control projects: Paradox Valley and Grand Valley, Colorado; Crystal Geyser, Utah; and Las

Vegas Wash, Nevada. It also formed the Colorado River Water Quality Improvement Program to investigate the feasibility of twelve additional salinity control projects. These projects have also been delayed and encountered large cost increases. The four projects identified above have experienced an estimated cost increase of 123 percent (from $125,100,000 to $278,775,000) since 1974. The twelve additional salinity control projects identified by WPRS have been subject to delays of from fifteen to sixty-four months between 1975 and 1977 and up to an additional 30 months according to an August 1978 schedule (U.S., GAO, Comptroller General 1979, pp. 28-31). Part of the reason for these delays is local opposition to the project. For example, at Paradox Valley there has been opposition to constructing large evaporative holding ponds. These delays are of considerable concern to the Salinity Control Forum, and may retard attainment of salinity goals for the river as water use intensifies by the year 2000.

Several other federal programs have been established for controlling pollution including salinity from agricultural lands and other nonpoint sources. One example is the 208 program of the states and EPA under the CWA. The responsibility for controlling salinity on federal lands belongs to the Bureau of Land Management of the DOI (CRB Salinity Control Forum 1978a, p. 55). Within the Department of Agriculture (USDA) there are several programs that deal with salinity. For example, the Soil Conservation Service is responsible for nonstructural activities such as improved land management to control salinity on agricultural lands. The USDA also administers the Rural Clean Water Program to correct salinity problems associated with "nonpoint" sources of pollution from rural (but not necessarily agricultural) lands (CRB Salinity Control Forum 1978a, pp. 54-55 and 78-79). Also, the Agricultural Research Service of the USDA in cooperation with WPRS and EPA is studying the Grand Valley area of Colorado to determine the effects of irrigation on salinity and is advising other agencies on practices which will reduce salinity from irrigation sources along the Colorado River. Another study being conducted by WPRS is designed to investigate saline water use and disposal opportunities in the Colorado River Basin (Hinds 1980). Further, the EPA Region VIII office has issued an energy policy statement which, among other provisions, supports the states' efforts in salinity control (U.S., EPA, Region VIII 1980).

In addition to these independent federal actions, each of the five CRB states has implemented its own water quality programs including salinity standards as set forth by the CRB Salinity Control Forum (1975) and revised in 1978. Specifically, the Salinity Control Forum

is currently preparing draft "baseline values" for salinity at twelve monitoring points throughout the Colorado
River system, identifying and evaluating river system
changes that may occur upstream of the monitoring sites,
overseeing the progress of salinity control projects,
and providing the member states with an overview of progress and problems in salinity control (CRB Salinity
Control Forum 1978a, pp. 27-29). Recently, the Salinity
Control Forum adopted a policy for industrial sources of
"no-salt return" whenever practicable and a limit to the
incremental increase in salinity from municipal discharges
in any portion of the river system of 400 mg/1 or less
(CRB Salinity Control Forum 1978a, pp.   A-1 through
A-10). The Salinity Control Forum has also recently
adopted a policy to encourage the use of brackish and/or
saline waters for industrial purposes (CRB Salinity
Control Forum 1980).

Although salinity standards at three points (Hoover,
Parker, and Imperial dams) and for industrial and municipal discharges have been agreed upon by the states in
the CRB and approved by EPA, it is unclear just how
these standards will be enforced and how the states will
respond to salinity increases. For example, even if the
salinity standards are exceeded at any of the three
points, there is no monitoring and enforcement mechanism
to determine the cause (or causes) of the salinity and
to control it. The baseline values being developed by
the states are intended to provide guidelines for monitoring salinity increases throughout the basin, but at
the present time these values are not intended as standards.

On the other hand, the states of the CRB Salinity
Control Forum feel the current standards and programs
are adequate to maintain salinity levels below the standards at least through 1990. The progress of water development and salinity changes in the Colorado River
system are monitored by the Forum. In the Forum's view
the adopted salinity control plan includes sufficient
measures to offset increases caused by levels of development estimated for 1990. The standards will thus be
met by implementation of salinity control projects, the
continuation of developments subject to effluent limitations that have been adopted by the basin states, and by
further development of management practices pursuant to
nonpoint contributions of salinity (CRB Salinity Control
Forum 1978a).

Due to the delay in federal salinity control projects, the basin states have adopted a policy of deeming
salinity increases above the 1972 levels caused by new
water developments to be in conformance with the standards if salinity control projects are underway which
would subsequently offset these "temporary" increases
(CRB Salinity Control Forum 1978a, p. iii).   The

Salinity Control Forum believes this policy is in conformance with federal regulations.

The slow progress towards achieving an effective salinity control plan in the CRB has been criticized by environmentalists, and the EDF filed suit against EPA for failure to comply with CWA requirements for stream standards, nondegradation of water quality, and development of an adequate control plan, including a compliance schedule covering streams in the Upper CRB. The EDF suggested that salinity standards be set for five additional stations, located at and upstream of Lees Ferry, to ensure control, monitoring, and guidance for future development. The Salinity Control Forum, however, argues that the present three stations and the anticipated baseline stations are adequate. The Forum does not consider state or subbasin standards to be the most cost-effective or consistent basin-wide approach to development of the river because, in part, it is more costly to clean up diffuse salinity sources in the Upper CRB than saltier streams in the Lower CRB (CRB Salinity Control Forum 1978b). The EDF lost in the district court and appealed to the Washington, D.C. Circuit Court of Appeals (Appeal no. 79-2432) (EDF 1980). In April 1981, the Appellate Court affirmed the District Court's opionion; EDF has not yet decided whether to appeal the case further (EDF 1981).

MUNICIPAL WASTEWATER POLLUTION

Summary: *Rapid relatively large population fluctuations in small western communities can cause water pollution due to the inadequate treatment of municipal sewage. Without assistance, few small communities in the West affected by energy development will be able to afford the cost of upgrading their capacities to meet new demands or of installing the secondary and tertiary treatment required by the CWA.*

The labor requirements of energy development can contribute to water quality problems in proportion to the size of the work forces required to construct and operate the facility. These labor requirements result in rapid and large population increases which impose heavy demands on the wastewater-treatment plants of small, western communities. Without proper advance planning and financial support, inadequate sewage treatment or even the bypassing of sewage treatment can occur. Our analysis and past experiences suggest that in many instances existing sewage-treatment plants are likely to be quickly overloaded by energy-related growth.

For example, an oil shale industry in western Colorado and eastern Utah with a 400,000 bbl/day capacity by 1990 would cause increased municipal wastewater of about 3 MMgpd during peak construction and 2 MMgpd during operation. Estimating capital costs for water supply and sewage treatment facilities at $1.76 million per 1,000 additional people, the capital costs for western communities, would be near $50 million in 1975 dollars.

Although effluents caused by population increases are much less than those associated with energy facilities, degradation of surface waters can still occur. The inability of most communities to afford upgraded facilities is complicated by the fact that more sewage treatment is needed during the construction phase of a facility than during its operation. Thus, it may be impractical to build sewage treatment plants to serve short-term peak demands only to have them underutilized later. In addition, insufficient sewer systems may affect other local problems; for example, new housing may be delayed and community health standards may be violated.

The CWA of 1977 requires municipal wastewater-treatment facilities to apply secondary treatment by July 1, 1978, and best practicable technology by July 1, 1984. Secondary treatment is currently defined by EPA as a biochemical oxygen demand not exceeding an average of 30 mg/l suspended solids not exceeding an average of 30 mg/l and a pH between 6.0 and 9.0. Best practicable technology is considered to be tertiary treatment that removes chemicals such as nitrates and phosphates.

Many energy-impacted communities in the West currently do not meet the secondary standards established by the CWA. Over one-half of the towns either lack a water or sewer system or have reached capacity in their existing systems (White et al. 1979c, p. 486). Upgrading facilities to meet the new standards will add to the financial burdens of local governments. Efforts to upgrade treatment facilities are aided by the Wastewater Construction Grants Program, which is administered by EPA through the states. However, various bottlenecks and complicated administrative procedures have been blamed for delays in meeting the 1977 deadline (U.S., CEQ 1975, pp. 71-72; Kirschten 1977b). The priority system established by the states for distributing EPA sewage treatment grants to nonmetropolitan communities places towns that have recently received funds far down on the list for future funds (depending on the specific state program). Water and sewer systems in towns where growth has continued, such as Gillette, Wyoming, are easily overburdened (Enzi 1977; Pernula 1977).

SUMMARY

Energy development activities increase conflicts over water quality by the disturbance of land, and the disposal of wastes. Extraction and upgrading or conversion of coal, oil shale, and uranium resources present the greatest risks to both surface water and groundwater quality. In addition, inadequately controlled exploratory drilling for oil, natural gas, and geothermal energy poses risks to groundwater quality by allowing contamination of fresh water aquifers with water from saline aquifers.

Land disturbance results from coal, oil shale, and uranium mining and constrution activities. Surface water quality is degraded by erosion of disturbed lands from blowing dust and rainfall runoff. This results in suspended and dissolved solids along with nutrients and some chemicals entering nearby streams. These chemicals from disturbed lands can also leach out, seep into aquifers, and reduce their water quality. Interception and disruption of groundwater aquifers during underground and surface mining of coal, oil shale, and uranium can degrade groundwater quality by allowing the mixing of fresh water aquifers with water from saline aquifers and by leaching of overburden and mine spoils into the aquifer.

Large volumes of wastes, some of which may be toxic to parts of the environment, are generated during oil shale, coal, and uranium mining, upgrading and conversion. The disposal of these wastes without adequate safeguards, such as within an aquifer or in an unlined evaporative holding pond, may eventually adversely affect surface water and groundwater quality. This occurs by seepage into groundwater or by berm failures which cause wastes to reach nearby streams. The CWA, Safe Drinking Water Act, RCRA, and the Toxic Substances Control Act address these problems, but the disposal and ultimate fate of wastes generated during mining and conversion of these resources are inadequately controlled.

Some of these laws and regulations may constrain energy development in the West. Regulations protecting groundwater quality mainly under the Safe Drinking Water Act and the SMCRA may be enforced to limit certain mining and waste disposal activities. Under RCRA, regulations are being formulated to control the disposal of several types of wastes from energy extraction and conversion. These regulations could substantially increase the cost of waste disposal.

Much political controversy is currently centered around salinity control in the CRB. Similar controversy may occur in the Yellowstone River Basin as energy resources are developed. In the CRB current salinity levels are substantially below the standards but salinity problems could constrain energy development if adequate salinity control projects are not built.

The effect of water consumption by energy facilities on salinity concentration levels highlights a weakness in the current policy system: the general lack within the states of a coordinated regulatory structure for water use and water quality. For example, water quality effects are generally not considered in a water use application. A second shortcoming of the current policy system of many states and the federal government is the inadequate predevelopment provision for municipal water and wastewater treatment facilities.

Several areas of inadequate technical knowledge are hampering the formulation of effective laws and regulations to protect water quality in some aspects of energy development. One of those areas is an inadequate knowledge of the groundwater resource base and its characteristics in energy resource areas. A second area of needed research is the development of better techniques to protect groundwater during mining and exploration activities. A third area of needed research is the characterization and classification of waste streams from various processes of energy conversion. For example, if a waste stream is considered hazardous, regulations under RCRA would substantially increase waste disposal costs. For emerging technologies such as coal liquefaction, unknown waste disposal costs can adversely affect their economic viability. Another example is the need for delineation of adequate waste disposal practices for spent shale.

Alternatives to deal with these problems are evaluated in chapters 4 through 6. Alternatives to utilize water more efficiently in energy conversion technologies and in agriculture are discussed in Chapter 4. In Chapter 5, use of saline water by energy facilities is discussed.

Several policy alternatives for protecting water quality are discussed in Chapter 6. These include predevelopment monitoring of groundwater resources and quality, the development of water quality control plans for individual facilities, and research into and monitoring of waste disposal practices. Alternatives to control salinity are also discussed including a salinity offset policy which ties water use with water quality and a policy of promoting salinity control projects. Temporary sewage treatment measures for energy-impacted communities are also discussed.

NOTES

[1]These estimates are based on U.S., DOI, BLM (1977, pp. III-135, III-136).

[2]For a description of solution mining of uranium see White et al. (1979a, vol. 4).

[3]At the salinity concentrations being dealt with here, milligrams per liter is essentially equal to parts per million.

Part II

Analysis of Policy Alternatives

4
Conservation of Water

INTRODUCTION

One approach to deal with both water availability and water quality issues is to reduce the demands for water while maintaining the same or similar levels of economic activity, that is, to conserve. In this chapter, water conservation includes reducing both the consumptive use of water and the water diverted from streams. Several alternatives for reducing water use in energy development and in irrigated agriculture are discussed.

Water use in energy development can be reduced in two basic ways: by modifying the process design to reduce raw water requirements and by using less water-intensive cooling technologies. For example, water consumption for coal conversion facilities can be reduced by using a combination of wet and dry cooling rather than the typical all-wet cooling system. Although not discussed in this chapter, policymakers can also reduce water consumption by influencing technological and locational choices. As discussed in Chapter 2, choosing to construct coal synfuel plants rather than coal-fired power plants in any particular area can have a substantial effect on water consumption. Likewise, using an in situ oil shale recovery process instead of above-ground retorting or locating a Lurgi coal-gasification plant in the Beulah area rather than in the Four Corners area reduces water needs. Although such locational and technological choices are not likely to represent an important region-wide strategy, they may be useful in reducing water conflicts in areas of greatest water scarcity.

As discussed in Chapter 2, irrigated agriculture is the largest consumer of water in the West. A reduction in the amount of water used in irrigated agriculture is possible with little adverse effect on crop quality or yield. Alternatives for reducing the consumption of water by irrigated agriculture include changing the method of irrigation and replacing water-intensive crops

such as alfalfa with vegetables, wheats, or other crops
which use less water per acre or less water per dollar
of crop value.

Conservation of water in municipalities is another
option for reducing water consumption in the West, but
this approach is not addressed in this report.   It
appears unlikely that it would be an effective region-
wide alternative because so few large municipalities
exist in the eight-state study area.   However, the
option can be important in some cases; for example,
reduced water demands in the Denver area could reduce
diversion requirements in rivers of the Upper Colorado
River Basin (CRB).

Each alternative will be described and then evaluated, based on the following criteria: potential for
water savings, water quality effects, economic cost,
environmental costs and benefits, and implementability.
The chapter will conclude with a brief summary and comparison of water conservation measures.

CONSERVATION IN THE ENERGY INDUSTRY

The amount of water consumed in energy conversion
facilities depends primarily on the type and location of
the energy conversion process, the process design, and
the type of cooling method.   This section discusses
methods to reduce water use in electric power generation
from coal, coal gasification, and coal liquefaction.
Water savings from modifications of the process design
of coal synfuel plants are discussed first.   Particular
attention is given to water savings by using different
means of plant cooling.

Process Design for Water Optimization

Process design decisions affect the fresh water
requirements, the amount of contaminated water to be
treated, and the amount of water for disposal.   The
figures used in our analysis of water requirements for
energy development assume a high degree of water use
minimization through process design optimization.   This
assumption reflects the fact that many water-minimizing
process designs are cost effective for the western situation, and many of the designs for plants proposed for
the West have included water minimizing processes.

In this section, process design considerations are
addressed only for coal synthetic fuel plants.   Although
only two examples will be discussed, the Lurgi gasification process and the Exxon Donor Solvent (EDS) liquefaction process, similar water savings measures can be

incorporated in other coal synfuel processes. Significant water savings from process design changes are not possible from coal-fired power plants and surface retorting of oil shale. For power plants, this is because the major water consumption step is cooling (see Figure 2-2). For oil shale, a large proportion of the water is used for disposing of spent shale; thus, there is little opportunity for large reductions in water consumption through process design changes.

WESCO Lurgi Coal Gasification Plant. The Western Gasification Company (WESCO) has proposed building four 250 million standard cubic feet per day (MMscfd) gasification plants in northwestern New Mexico (WESCO 1974). Since water supply is limited in this region of New Mexico, water conservation is important to this development. The principal consumptive uses of water in a Lurgi plant are for process and plant cooling, hydrogen supply, steam generation, sulfur pelletization, ash quenching, and road wetting. WESCO has examined each of these uses and has tried to maximize the reuse of wastewater from one process to supply another. This has been accomplished by incorporating reuse in the following ways:

River water will be used only for steam generation, sulfur scrubbers, and domestic plant water, with steam generation comprising the major fraction;

A portion of the steam is used by turbines which run compressors. The exhaust steam from the turbines will be condensed using dry cooling towers, recovered and reused for steam generation;

A major portion of the steam is used in the Lurgi gasifier both as a source of hydrogen and also as a temperature modulator. Some of this steam can be recovered, after special treatment, to supply water for plant cooling;

Blowdown from the cooling water will be used to quench the hot coal ashes from the Lurgi gasifier;

Mechanical refrigeration will be used where possible to replace wet cooling systems where dry cooling would be inadequate, thus eliminating a water requirement; and

Boiler blowdown water and wastewater from the necessary demineralizers, along with treated sewage and all other recoverable water, will be used for sulfur pelletization, wetting roads for dust control, and for mining and coal processing operations.

The result of incorporating these water saving measures is a total consumptive water requirement of 7,620 acre-feet per year (AFY). As shown in Table 4-1, 11 percent of the total requirement is used for process consumption, 72 percent is returned to the atmosphere, 9 percent is disposed of at the mine, and 8 percent is used for miscellaneous purposes.

Exxon Liquefaction Process. Process design for water use optimization has also been studied in the EDS coal liquefaction process (Exxon 1977, p. 97). In a design study for a commercial plant located at a Wyoming site, a reduction in water requirements of 72 percent was achieved by three process design changes. The first step is to examine phases in the conversion process that can be changed to save water in a cost effective way. Cost effectiveness is used to describe modifications which are, at least, marginally economic even when the objective is not to reduce overall water consumption. One such area is to use effluent water for ash handling makeup water. Nine of these phases in the process accounted for 21 percent of the total reduction of 72 percent.

The second step in process design involves the method of producing the hydrogen. Partial oxidation is used in the Exxon plant instead of steam reforming. As a result of the change, treated effluent water can be used for hydrogen production instead of high quality water. This also reduces wastewater disposal problems. Although Exxon is still studying problems with reliability and the degree of development required, they indicate that this change, accounting for 15 percent of the 72 percent reduction, looks economically attractive.

The third step in process design is the use of treated plant effluent water to supply cooling tower requirements; that is, the water used to replace the quantity evaporated in the cooling process. Problems with cooling towers, such as biological growth and scaling, are increased with this method but are partially mitigated by chemical treatment of the cooling water. This step, accounting for half of the 72-percent decrease in water requirements, "is likely to become a viable alternative by the time a commerical plant is built, based on

TABLE 4-1:   Water Requirements and Disposition:
             WESCO Gasification Plant

	AFY	Percentage
Process consumption		
To supply hydrogen	1,800	
Produced as methanation by-product	-960	
Net consumption	840	11.0
Return to atmosphere		
Evaporation:		
From raw water ponds	100	
From cooling tower	2,870	
From quenching hot ash	250	
From pelletizing sulfur	400	
From wetting of mine roads	1,500	
	5,120	
Via stack gases:[a]		
From steam blowing of boiler tubes	320	
From stack gas SO_2 scrubbers	60	
	380	
Total return to atmosphere	5,500	72.2
Disposal in mine		
In water treating sludges	160	
In wetted boiler ash	50	
In wetted gasifier ash	480	
Total disposal in mine	690	9.1
Others		
Retained in slurry pond	35	
Miscellaneous mine uses[b]	555	
Total others	590	7.7
Grand Total	7,620	100.0

Source: Compiled from WESCO 1974.

[a]Does not include water derived from burning of boiler fuel

[b]Primarily selective irrigation

expected industry development of the required technology" (Exxon 1977, p. 99).

Evaluation. The WESCO and Exxon examples are complicated processes which have been only briefly described to illustrate potential water reductions from process design changes. This reduction is important in the West for technical, political, and economic reasons. For example, in the WESCO case, "the total raw water available to WESCO from the San Juan River is contractually limited to 44,000 AFY by the U.S. Department of the Interior." Since four gasification plants are planned by WESCO, water minimization is mandatory for both water conservation and environmental quality. Based on present design, the first plant will require 7,620 AFY, which provides "a safety margin for unforeseen contingencies" (WESCO 1974, pp. 12-13).

A second advantage of process design change is the reduction in wastewater disposal requirements. For example, Step 1 modifications for the Exxon case decrease the plant sour water treating load by approximately one-half of the base case rate (Exxon 1977, p. 99). In fact, one of the prime justifications for the Exxon study was to reduce the investment and operating costs to treat wastewater.

The only potentially negative impact of water use minimizing design is an increase in the cost of the product. However, current indications are that such changes either save money or only increase costs slightly. A wide range of possible process design changes exist, and the economic effects will depend on a variety of site-specific factors, such as raw water cost. However, Exxon believes that only minor price increases will result from this alternative (Exxon 1977, p. 99).

Cooling Technologies for Energy Conversion Plants

Water requirements for energy conversion plants may also be reduced by employing less water-intensive cooling technologies. If high wet cooling is used, cooling accounts for 65 to 90 percent of the net water consumed by the energy conversion facility. This percentage varies with the type of energy conversion facility. In this section, a dry cooling option and combinations of wet and dry cooling are discussed.[1] Power plants are considered first, followed by coal synfuel plants. Dry or wet/dry cooling could also be used in oil shale retorting processes but neither is considered in this study.

Coal-Fired Steam Power Plants. In a steam power plant, coal is burned to produce high temperature, high pressure steam. Some of the heat is converted to work in the turbines which turn generators to produce electricity. This process is typically about 38 percent efficient in new plants. The remaining 62 percent of the heat is unrecoverable and must be cooled. Wet cooling evaporates the water in cooling towers as a means of transferring heat to the atmosphere. In dry cooling systems, the waste heat is transferred to the atmosphere by "sensible heat transfer." Because air is a poor heat-transfer medium compared to water, dry cooling of plants necessitates larger air volumes, larger heat transfer surfaces, and larger towers than evaporative water systems. This means that dry cooling systems have a larger capital cost; a plant with a dry cooling system costs roughly 10 to 15 percent more than a plant using a wet cooling system.

Dry cooling can also lower the efficiency and capacity of the facility, depending on how it is designed. The fans which create the draft in dry cooling towers have large energy requirements. Additionally, a dry cooling system can only cool to temperatures approaching that of outside air. Thus, on a very hot day, temperatures of the coolant water are high, creating high back pressures on the turbines, reducing conversion efficiency, and causing lowered output capacity. To account for reduced efficiency and capacity, the plant may have to be enlarged to meet the design output. For this reason, it is often desirable to have wet cooling towers available for hot summer days. The option of total dry cooling is typically only considered in winter peaking utilities, where peaks in demand correlate with low ambient air temperatures. Although economics usually preclude its use in the West, it is currently employed in the 330 MWe plant at Wyodak, Wyoming. Because of the high cost of total dry cooling and its limited practice in the study area, emphasis is placed on wet/dry cooling in this chapter.

Table 4-2 summarizes water consumption for wet and wet/dry cooling for five sites in the study area (Gold and Goldstein 1979). The range in the water consumed for both wet and wet/dry cooling is due to site-specific factors; the plants located in warmer regions require more wet cooling. A considerable water saving is possible using wet/dry cooling. Cooling water requirements can be reduced by about 85 percent, resulting in a total reduction in the water use of the plant and associated mines of 68 to 77 percent. The 3,000 MWe steam power plants with wet/dry cooling have the same range of water requirements as the "standard size" synfuel plants with all wet cooling (see Chapter 2 and Table 4-3). The cost of achieving these water savings is discussed below.

TABLE 4-2:   A Comparison of Wet and Wet/Dry Cooling
for Coal-Fired Power Plants

Technology	Site	Net Water Consumed (AFY)		Water Saved (AFY)	Percent Savings
		Wet Cooling	Wet/Dry Cooling		
Steam Power	Beulah	23,900	5,500	18,400	77
Generation	Colstrip	26,700	7,300	19,400	73
(3,000 MWe)[a]	Gillette	25,800	6,500	19,300	75
	Farmington	29,200	9,100	20,100	69
	Kaiparowitz	29,800	9,500	20,300	68
	Range	23,900–29,800	5,500–9,500	18,400–20,300	68-77

Source: Gold and Goldstein 1979, Table 1-3, p. 10; Gold et al.
1977.

[a]70 percent load factor.

Coal Synfuel Plants.   Coal synfuel plants have
several phases in the conversion process that require
the use of cooling water. For example, compressors are
used to transfer various gases from one phase of the
conversion process to another; these compressors are
driven by steam turbines and the exhaust steam from
these turbines must be cooled. Additionally, some of
the intermediate process gases must be cooled. In using
dry cooling for one phase and wet cooling for another,
the advantages of each cooling process can be exploited.
The following discussion summarizes a study that examined three cooling options for coal synfuel plants:
high wet, intermediate wet, and minimum practical wet
cooling (Gold and Goldstein 1979). High wet cooling is
mostly wet cooling, but due to process design optimization, some dry cooling has been incorporated for cooling
very high temperature gas streams. For intermediate wet
cooling, dry cooling is substituted to handle about 90
percent of the load on the turbine condensers. The minimum practical wet cooling case adds dry cooling to handle 50 percent of the load on the interstage compressors

The consumptive water requirements for the different cooling technologies for four sites in the West are summarized in Table 4-3. Net water savings do not differ substantially among these technologies and locations; with intermediate wet cooling, water savings range from 1,500 to 2,100 AFY. Among these technologies, Synthoil liquefaction is the most water-intensive and has the smallest percentage of decrease in water consumption by employing water saving cooling technologies. Among the cooling options, minimum practical wet cooling provides little additional saving over intermediate wet cooling; only an additional 4 to 10 percent saving is achieved. In contrast, 16 to 32 percent saving is obtained by using intermediate wet cooling rather than high wet cooling. As discussed in the following section, these comparisons are important to the economics of cooling technologies.

Evaluation. One important question about the use of water saving technologies is economics. The economic trade-off is the increased capital and operating costs of dry cooling systems compared to the cost of water. Water costs include the price of the water at its source, transporting it to the site, and water treatment. If water is free, the obvious choice is wet cooling. This is almost never the case in the study area; thus water conservation technologies can often be economical.

"Break-even" water costs have been calculated for the three cooling options, three coal conversion technologies, and four sites (Gold and Goldstein 1979, Chapter 4). These break-even, or change-over, costs are calculated by dividing the increased annualized plant cost for the dry or wet/dry system, assuming water costs are negligible, by the water saving obtained. The results of these calculations are summarized in Table 4-4. Since the break-even costs do not vary greatly among the Lurgi, Synthoil, and Synthane processes, they are combined in a single coal synthetic fuel category.

For coal synfuels, wet cooling is the economical choice if water costs less than $35 per AF.[2] For costs between $35 and $81 per AF the economics are such that the choice between high wet cooling and intermediate wet cooling must be made on a case-specific basis. For costs above $80 and up to $300 per AF, intermediate wet cooling is the most economical. The "break-even range" between intermediate wet and minimum practical wet cooling for synfuel plants is between $300 and $490 per AF. The actual cooling technology employed would be determined by many case-specific factors for the energy facility. For water costs above $500 per AF, minimum wet cooling technologies are most economical.

TABLE 4-3: Water Savings for Coal Synfuel Technologies[a]

Technology[b]	High Wet Cooling	Intermediate Wet Cooling			Minimum Practical Wet Cooling		
	Net Water Consumed (AFY)	Net Water Consumed (AFY)	Net Saving (AFY)	Saving as % of Wet Cooling	Net Water Consumed (AFY)	Net Saving (AFY)	Saving as % of Wet Cooling
Lurgi (250 MMscfd)	4,900-7,100	3,300-5,600	1,500-1,600	21-32	2,900-5,200	1,800-2,100	27-42
Synthane (250 MMscfd)	7,700-8,700	5,900-6,700	1,800-2,000	23-24	5,500-6,300	2,100-2,400	27-29
Synthoil (100,000 bbl/day)	9,200-11,800	7,500-9,700	1,700-2,100	16-19	7,000-9,100	2,200-2,700	21-25

bbl/day = barrel(s) per day

Source: Gold and Goldstein 1979, Table I-1.

[a]Range shown is across four sites: Navajo/Farmington area in New Mexico; Gillette, Wyoming; Colstrip, Montana; and Beulah, North Dakota. A relatively high degree of water minimization through process design is also incorporated in these figures.

[b]90 percent load factor.

TABLE 4-4:   Break-Even Water Costs
              (dollars per AF)[a]

Technology	Intermediate Wet	Minimum Wet
Power Plants	$1,200-$1,900	N.C.
Coal Synthetic Fuels	$35-$81	$300-$490

NC - not considered

[a]Range is over the sites considered and for a range of unit prices
considered (e.g., fuel cost and the annual fixed charge rate).

For steam electric generation, wet cooling is still
economical up to $1,200 per AF. For water costing be-
tween $1,200 and $1,900 per AF the trade-off between all
wet and wet/dry cooling depends on site-specific factors. If water costs above $1,900, wet/dry cooling of
steam electric generation is the economic choice.

The break-even water costs for power plants are
higher than for synthetic fuel plants because dry cooling equipment for power plants necessitates higher capital costs. The main difference is in the design of the
cooling tower and condenser systems. In power plants,
"dry cooling" means a wet condensing system and a dry
cooling tower. In synthetic fuel plants, which are
similar in operation to refineries, there are several
smaller condensers which are dry cooled. This elimi-
nates the costs of the wet condensers, the circulating
water system, and the dry cooling tower.

Although the costs of water rights are often relatively small,[3] the cost to transport water to a given
site may not be. Except for energy facilities located
near the main stem of major rivers or near large reservoirs, the cost of transporting water is in excess of
$81 per AF, indicating that intermediate or minimum
practical wet cooling is economically desirable for most
sites in the study area. However, for power plants,
high wet cooling is usually the most economic choice
since the cost of purchasing, transporting, and treating
the water, and disposing of the cooling tower blowdown
is generally much less than $1,200 per AF.

If energy developers are forced to use water saving
cooling technologies, product costs will increase, as
indicated in Table 4-5. These cost increases are based
on the assumption that water costs are negligible, and
therefore they represent the maximum economic penalty
that could be effected. The costs of adopting intermediate wet cooling for synfuel plants are almost

TABLE 4-5: Maximum Economic Costs of Cooling Water Savings[a]

Technology	Changing from "High Wet" to "Intermediate Wet" Cooling		Changing from "High Wet" to "Minimum Wet" Cooling	
	Water Saved (AFY)	Cost	Water Saved (AFY)	Cost
Power Plant (cents/kWh)	18,400-20,000	0.11-0.18	NC	NC
Lurgi Gasification (cents/million Btu)	1,500-1,600	0.15-1.03	1,900-2,100	1.18-1.32
Synthane Gasification (cents/million Btu)	1,800-2,000	0.18-1.22	2,100-2,400	1.33-1.47
Synthoil Liquefaction (cents/million Btu)	1,700-2,100	0.14-1.00	2,200-2,700	1.06-1.27

[a]These calculations assume the cost of water is negligible (see text). Estimates are derived from Gold and Goldstein (1979). Ranges reflect site-specific differences.

negligible. For example, if the cost of synthetic natural gas is approximately $4.50 per million British thermal units (Btu), in the worst case for the Lurgi process, product costs would increase less than .23 percent due to intermediate wet cooling. This nominal price increase would save 1,500 to 1,600 AF of water per year. In contrast, the cost of conserving water in steam-electric generation plants is considerably higher. Assuming it costs 4.0 cents per kilowatt-hour (kWh) to generate electricity in new steam-electric plants (excluding transmission and distribution costs), electricity costs would increase 2.75 to 4.5 percent. While the price is higher for steam-electric plants than for synfuel plants, the water saving is quite large. For a typical plant in the study area the water saving would be about 20,000 AFY.

The economic penalty of adopting the minimum practical wet cooling technology in synfuel plants is also not large. However, it is high relative to the percentage water savings. That is, use of minimum practical wet cooling rather than intermediate wet saves about 25 to 30 percent of the amount of water saved in going from high wet to intermediate wet cooling, but the cost is almost ten times as much. Although this marginal cost of minimum wet cooling is high, it will still add only about 0.3 percent to the total cost of the synthetic fuel.

The costs presented above are maximum penalties. Any cost incurred in acquiring, transporting or treating the water would decrease the actual economic penalty. Therefore, it appears that intermediate wet cooling practices could be economically employed at all coal synfuel plants in the study area. For steam-electric power plants, if only the costs of water are considered, then all wet cooling will usually be desired. However, because water is in short supply, and there are many pressures for conserving water, wet/dry cooling can be desirable in many western locations.

Environmentally, conservation of water in energy conversion facilities will have a positive effect if water saved remains in the stream rather than being withdrawn for use. While water savings per facility are not large, they could have a significant effect on streams with low flows or on a basin-wide scale if many conversion facilities are sited. However, in the case of steam-electric power plants, care must be exercised in interpreting this result. Although total water consumption is reduced by 67 to 75 percent, the largest water needs for wet/dry systems occur during the summer, when streamflows are generally lowest. For example, for Farmington, New Mexico, a wet/dry tower consuming 10 percent of the annual water consumed by an all-wet tower will have a maximum flow rate of almost 60 percent of that required by an all-wet system (Hu and Englesson 1977). Therefore, unless sufficiently large water storage impoundments are used, instream flow conditions could still be adversely affected when wet/dry systems are utilized in power plants.

A potentially negative environmental effect of water conservation is related to the possibility of lowered energy conversion efficiencies. If the plant efficiency is lowered, then for the same amount of output, more input resources would be required. This in turn would increase land impacts from mining, air pollution, and solid waste disposal rates. However, for coal synfuel plants, no significant decline in energy conversion efficiency is expected either for process changes or for wet/dry cooling. For power plants the use of dry cooling towers can lower conversion efficiencies due to the fan requirements and higher condenser temperatures on hot days. However, when wet/dry cooling is used and the plants are "optimally" designed, average power plant efficiencies are not expected to be significantly lowered (Gold 1980).

A critical question about this alternative is how it would be implemented. In general, knowledge about locational variations in water problems and in the influence of new technologies suggests that implementation procedures will need to be flexible enough to allow states and localities to meet particular demands and characteristics of specific areas. For example, if a policy requiring "best available water conservation technology" were uniformly applied, it would negate knowledge about the importance of locational impacts in determining water availability problems and issues. Indeed, water conservation is likely to be needed in different proportions in various areas of the West, depending on local demands and water quantities available.

Flexibility also includes the need for state and local decisionmaking responsibility, which was highlighted by the conflicts between the western states and the Carter Administration regarding federal water policy. Federal attempts to mandate uniform conservation policies which do not include strong state roles in formulating and implementing the policies are likely to be very difficult to enact. This issue is discussed in more detail in Chapter 8.

ALTERNATIVES FOR IMPROVING WATER USE IN AGRICULTURE

Irrigated Agriculture Water Use

Water availability and water quality can also be enhanced by improving the use of water in the agricultural sector. As discussed in Chapter 2, irrigated agriculture is the largest diverter and consumer of water in the study area, and it is a major component of the agricultural productivity of each state in the study area, except North Dakota.

As shown in Table 4-6, considerable portions of cropland in Arizona, Colorado, New Mexico, Utah, and Wyoming are irrigated. The percentage is highest in Arizona, where 95 percent of cropland is irrigated. In the remaining four Rocky Mountain states, the percentage of irrigated cropland ranges from 40 to 75 percent. In each of these 5 states, irrigation is the largest single water consumer. Since most of the annual precipitation in the Northern Great Plains occurs during the growing season, much less supplemental irrigation is required in Montana (16%) and North Dakota (1%). Excluding North Dakota, the amount of water required per acre varies considerably across the remaining six states in Table 4-6; from 1.7 acre-feet per acre (Montana) to 4.3 acre-feet per acre (Arizona). The predominant methods of irrigation in these states are flooding and furrows or ditches. Each state has some acreage irrigated by sprinkler systems.

The remainder of this section evaluates two alternatives for reducing water consumption in agriculture: improved irrigation techniques and crop switching. In addition to evaluating potential water savings, water quality implications are considered.

Conservation Through Improved Irrigation Practices

Improved irrigation practices and their effect on water use and quality are largely determined by irrigation "efficiency." Irrigation efficiency is affected by both off-farm conveyance systems (unlined canals, lined canals, or pipelines) and by on-farm conveyance and application systems.

In the case of off-farm conveyance systems, efficiency is the volume of water delivered to the farm as a percentage of the total volume of water diverted (gross diversion) from a stream or other water supply. On-farm efficiency refers to the volume of water necessary for adequate crop growth (crop consumptive use) as a percentage of the volume of water delivered to the farm. Overall irrigation efficiency thus is the product of off-farm and on-farm efficiencies; that is, the total water consumed by crops expressed as a ratio or percentage of the total water diverted from a water supply. For example, if a farmer diverted 100 AF of water from a stream and delivered 75 AF of water to the farm, the off-farm irrigation efficiency value would be 75 percent, with a 25 percent conveyance loss. If the amount of water necessary for crop growth on a parcel of land is

TABLE 4-6: Profile of Irrigated Farms Having Sales of $2,500 or Over

Category	Arizona	Colorado	Montana	New Mexico	North Dakota	Utah	Wyoming
Average Annual Rainfall (inches)	14	17	15	15	17	13	14
Total Cropland (acres)	1,372,631	10,207,361	15,257,374	1,900,964	28,997,608	1,668,708	2,555,196
Cropland Harvested (acres)	1,052,199	5,892,090	8,375,116	926,931	19,158,983	1,024,528	1,634,395
Total Pasture Land, all types (acres)	15,033,878	23,517,413	41,177,450	35,997,895	11,767,559	7,093,061	27,840,922
Number of Farms Irrigated	2,984	12,324	7,528	3,975	490	7,281	4,195
Percent of Farms	69.1	58.5	37.3	51.4	1.2	83.3	60.9
Total Land Irrigated (acres)	1,109,978	2,788,746	1,717,764	816,754	69,713	897,192	1,426,725
(acres/farm)	372	226	228	204	142	123	340
(percent of cropland)	80.9	27.3	11.3	43.0	0.24	53.8	55.8
Cropland Irrigated (acres)	1,048,819	2,641,159	1,606,798	785,674	68,160	840,085	1,255,062
Harvested Cropland Irrigated (acres)	998,802	2,357,978	1,367,432	693,570	66,341	707,678	1,059,496
Percent of Total Cropland Harvested	94.9	40.0	16.3	74.8	0.35	69.1	64.0
Cropland Pasture Irrigated (acres)	31,172	252,691	226,404	57,822	1,375	123,785	190,732
Other Croplands Irrigated (acres)	18,845	30,490	12,962	34,282	494	8,622	4,832
Improved Pasture Land and Range Land Irrigated (acres)	11,142	147,587	110,966	25,080	1,553	57,107	171,663
Irrigated Water Applied	4,779,771	5,518,212	2,985,625	1,898,398	90,084	2,041,662	2,754,152
(acre-feet per acre)	4.3	2.0	1.7	2.3	1.3	2.3	1.9
Method of Irrigation							
Furrows or Ditches (acres)	754,091	1,354,121	532,523	478,131	22,766	367,249	572,568
Flooding	309,849	1,036,979	906,762	189,900	28,783	420,813	738,347
Self-Propelled Sprinkler Systems (acres)	18,809	304,785	61,581	91,400	14,873	11,688	58,679
Other Sprinkler Systems (acres)	20,481	91,771	209,298	58,481	4,236	103,465	49,197
Subirrigation (acres)	18,419	49,597	50,894	6,551	545	13,410	34,284

Source: U.S., Dept. of Commerce, Bur. of Census 1974.

30 AF, the on-farm irrigation efficiency would be 40 percent. Thus, the overall efficiency in this situation is 30 percent (.40 x .75).

Irrigation efficiencies in the United States average 41 percent--78 percent for off-farm conveyance efficiencies and 53 percent for on-farm efficiencies (Interagency Task Force on Irrigation Efficiencies 1979). As shown in Table 4-7 and Figures 4-1 and 4-2, in the eight-state study area overall irrigation efficiencies average 34 percent, with on-farm efficiencies of 47 percent and off-farm efficiencies of 72 percent (USDA, SCS, Special Projects Div. 1976). Several studies have concluded that water diversions can be reduced and water quality improved by increasing these efficiencies (Interagency Task Force on Irrigation Efficiencies 1979, p. 3; Utah State U., Utah Water Research Lab. 1975, pt. 2, pp. 249-63; USDA, SCS, Special Projects Div. 1976, pp. 13-16; U.S., DOI, BuRec and BIA 1978, pp. x-xi). Furthermore, these water savings would generally not necessitate a reduction in crop production.

Description of Alternatives

Whether the primary objective is improved water quality or reduction in water consumption, the techniques used to increase efficiency would be the same. Off-farm irrigation improvements include ditch lining, piping, removal of noncrop vegetation along ditches, and flow regulating and monitoring systems. Ditch lining or piping is a method of reducing water loss due to seepage or evaporation. Ditches can be lined with impervious substances, such as concrete or asphalt, or they can be replaced by pipelines. The reduction in percolation and evaporation will reduce soil leaching and thus salinity problems and nutrient loss. Water measuring devices are necessary to accurately account for the quantity of water used and where it goes in the system. Flow regulators reduce operational waste by allowing only the water that is needed into the system.

On-farm improvements include land leveling, on-farm ditch lining or piping, automation of the irrigation system, and more efficient application methods. Ditch lining, automation, and land leveling are changes which can improve efficiency in surface irrigation systems (i.e., flooding and furrow or ditch systems).

Three irrigation application methods, surface, sprinkler, and trickle/drip, are generally accepted as being adequate for most irrigation situations. Since the efficiency of each method depends on area characteristics, no single method can be designated as the most efficient. For example, in some areas of the West, where slope and coarse-textured soils are characteristic,

124

TABLE 4-7: 1975 Irrigation Efficiency

	On-Farm	Off-Farm	Overall
Arizona	56%	78%	44%
Colorado	46%	71%	33%
Montana	42%	50%	21%
New Mexico	50%	80%	40%
North Dakota	50%	67%	34%
South Dakota	50%	86%	43%
Utah	44%	78%	34%
Wyoming	35%	69%	24%
Average for study area	47%	72%	34%
Average U.S.	52%	78%	41%

Source: Derived from USDA, SCS, Special Projects Div. 1976.

it may be appropriate to change from a surface irrigation technique to sprinkler irrigation. However, in
areas where climatic conditions are such that an excessive amount of water would be lost with sprinkler irrigation due to evaporation, or where salt concentrations
in the soil are high and need to be leached out, surface
application remains appropriate.[4] In areas where the
water supply is extremely limited trickle irrigation may
be the most appropriate (Interagency Task Force on
Irrigation Efficiencies 1979, pp. 73-81). Off-farm and
on-farm alternatives are evaluated according to total
water savings, water quality, economic costs, and implementation.

Water Availability. Uncertainty exists about the
amount of water which could be saved and therefore used
for other purposes by improvements in on-farm and off-
farm efficiencies. However, in general, decreasing the
amount of water diverted relative to crop consumption
will not substantially increase the supply of water to
downstream users. In a typical irrigation water budget
for the eight-state study area (see Figure 4-3), approximately 41 percent of the total water diverted is
consumed by the crops being irrigated. This total cannot be reduced unless the crop type, acres irrigated, or
cropping pattern (e.g., contour plowing) is changed.

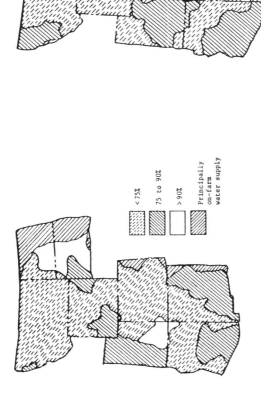

45% or less

46 to 65%

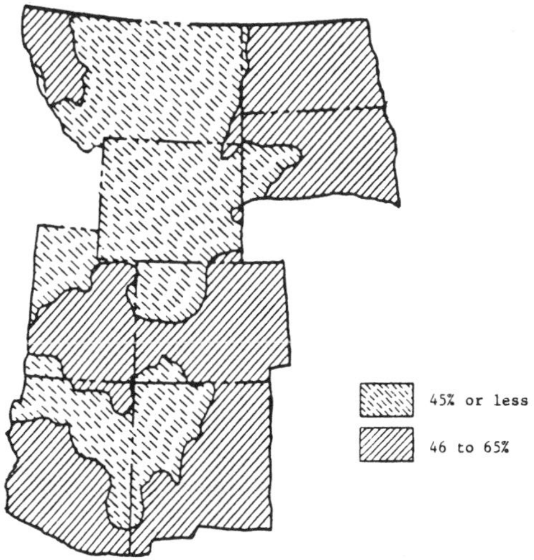

Figure 4-2: Average On-Farm Efficiency Within a Drainage Basin

Source: Interagency Task Force on Irrigation Efficiencies 1979, p. 25.

< 75%

75 to 90%

> 90%

Principally on-farm water supply

Figure 4-1: Average Off Farm Efficiency Within a Drainage Basin

Source: Interagency Task Force on Irrigation Efficiencies 1979, p. 24.

126

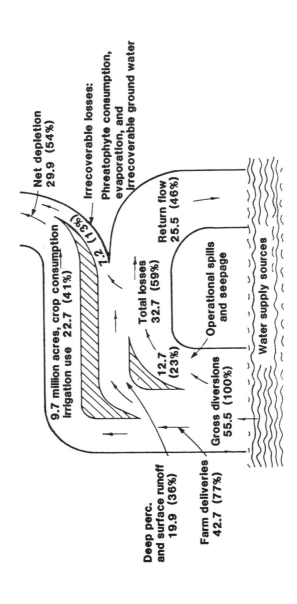

Figure 4-3: Irrigation Water Budget for the Eight-State Study Area with Normal Year Water Supply, 1975 Level of Development (quantities in millions of acre-feet).

Adapted from: USDA, SCS, Special Projects Div., 1976.

The difference between crop consumption and total diversion is related to return flow and irrecoverable losses. Return flow is the portion of water diverted for irrigation which returns to accessible surface or groundwater. Return flow typically amounts to 46 percent of total diversion (Interagency Task Force on Irrigation Efficiencies 1979). Irrecoverable losses refer to that amount of water diverted which is lost and unavailable for further use. These losses include phreatophyte consumption, evaporation, and loss to irretrievable groundwater. Irrecoverable losses will vary by region and irrigation practices; average estimates of irrecoverable losses range from negligible to 13 percent (Gilliland and Fenner 1979; U.S., Water Resources Council 1978; Interagency Task Force on Irrigation Efficiencies 1979). However, in some basins this amount could be as high as 20 percent (Interagency Task Force on Irrigation Efficiencies 1979).

For the eight-state study area, irrecoverable losses have been estimated at 13 percent or 7.2 million acrefeet per year (MMAFY) as shown in Figure 4-3. Water consumption in agriculture can be reduced only through a reduction in irrecoverable losses with a given crop type, irrigated acreage, and cropping pattern. Because much of the irrecoverable water loss is due to evaporation and percolation from the surface area, practices to reduce irrecoverable loss have limited effectiveness. For example, although sprinkler irrigation reduces the amount of water diverted, evaporation can actually increase depending on climatic conditions (Gilliland 1981). Lining canals also has limited effectiveness because much of the excess water diverted, but not applied to crops, returns from the canal margins to adjacent streams. Canal lining serves mainly to reduce diversion requirements and control evapotranspiration losses by phreatophytes which represent less than 10 percent of irrecoverable water loss (Gilliland and Fenner 1979). As discussed later, sprinkler irrigation and canal lining are also quite expensive. It is doubtful, then, that significant portions of the 7.2 MMAFY irrecoverable losses can be managed to reduce net depletions by agriculture. In addition, changes in irrigation practices to reduce depletions may affect the pattern of return flows from agricultural lands. Because adjacent irrigators have a claim to these return flows, opportunities to change consumptive use may be limited and water apparently made available by reduced diversions may not be reallocated because downstream users may depend on return flow. Thus, the only water available for other consumptive use is the amount freed by a reduction in incidental losses (Pring and Tomb 1979, p. 9; Interagency Task Force on Irrigation Efficiencies 1979; U.S., DOI, BuRec and BIA 1979, p. xi).

However, even if net depletions are not reduced, improved irrigation efficiencies could decrease gross diversions. For the CRB, the use of sprinkler irrigation to improve efficiency to the 80 percent level would reduce diversions by 1.8 acre-feet per acre (from 4.9 acre-feet per acre to 3.1 acre-feet per acre)--a 36 percent reduction. This can have important implications for water quality as discussed below (Utah State U., Utah Water Research Lab. 1975).

Water Quality. Improving irrigation efficiency will probably be more important for water quality. Reductions in the amount of water diverted and applied to the land would subsequently decrease amounts of return flow. The total amount of salts, pesticides, and other pollutants in return flow decreases as a result of less water percolating through the soil. However, water percolation removes accumulated soil salts and can be important in some locations to sustain crop yield (Interagency Task Force on Irrigation Efficiencies, Technical Work Group 1978, p. 17). It has been estimated that agricultural return flows contribute 1.5 million tons of salt per year to the Upper CRB at Lees Ferry and contribute approximately 5.7 million tons of salt annually to the Upper and Lower CRB.

As of 1975, irrigation efficiencies for the sub-basins in the Colorado River system were estimated to average 48 percent, including 80 percent off-farm efficiency and 61 percent on-farm efficiency (Utah State U., Utah Water Research Lab. 1975, pt. 2, pp. 146-65). If irrigation efficiencies were increased to 80 percent, diversions would be reduced an average of 1.8 acre feet per acre. Assuming the water saved remained in the stream, salinity concentrations and loading from return flow would be reduced. At 80 percent efficiency in off-farm and on-farm systems as projected in the Colorado River Regional Assessment Study, salt content at Imperial Dam could be reduced by 1.96 million tons per year (MMtpy) from 1977 to 1985. Estimates of changes in salinity concentration at Imperial Dam for this period were 136 milligrams per liter (mg/l) for off-farm improvements (canal lining) and 35 mg/l for on-farm improvements (sprinkler irrigation).

In short, improvements in irrigation efficiencies can effectively control salinity and other water pollution problems associated with agricultural return flows. Chapter 6 compares this to several other approaches to salinity control.

Economic Costs. Generally, both on-farm and off-farm improvement costs are high. Costs will vary by crop type, acreage cultivated, irrigation practices, and local and seasonal climatic conditions. Estimates of the costs of converting to sprinkler irrigation in the CRB are from $23 to $127 per acre per year with an average of about $50 per acre per year (1975 dollars) (Utah State U., Utah Water Research Lab. 1975, pt. 2, p. 249). Assuming this cost and an estimated reduction in diversion of 1.8 AF per acre, a one AF reduction in diversion would cost about $28 per acre per year. Of course, costs vary considerably, depending on the area. In southern Arizona the total annualized cost per acre for installation of sprinkler irrigation systems could range from $64 to $148 (1977 dollars) depending on the total acreage irrigated (Wade et al. 1977). A sprinkler system for a typical alfalfa farm (4,000 acres) in the San Juan River Basin in New Mexico would increase total annualized costs by about $40 per acre (Gilliland and Fenner 1979, p. 4-16).

The estimated increase in annual costs from a conversion to sprinkler irrigation is due in part to fuel and power costs associated with pumping. The costs of pumping water at a pressure sufficient to operate the sprinklers must be considered. In addition, variations in energy costs result in different cost estimates by region and by fuel source. An estimate of the average per-acre power cost for sprinkler irrigation in the CRB is $30 (Keller, as cited in Utah State U., Utah Water Research Lab. 1975, pt. 2, p. 247). This estimate is based on 1975 electric power rates for the area. In some areas pumps can be powered by other methods, such as diesel fuel. These variations in costs should be considered when total annual cost increases for sprinkler irrigation technologies are assessed.

Canal lining is also costly. Estimates for the CRB in 1975 were $30,000 to $40,000 (1975 dollars) per mile of canal lined. Canal lining costs per acre for the basin ranged from $72 to over $600 for the Upper CRB with an average of $250 per acre (Utah State U., Utah Water Research Lab. 1975, pt. 2, p. 264). In the San Juan River Basin in New Mexico estimates for canal lining were reported to be $80,000 to $250,000 per mile lined, depending on the width of the canal (Gilliland and Fenner 1979, p. 4-13). In the average case, initial capital investments would average $600 per acre, increasing annual costs by $50 to $60 per acre.

Depending on the price of water and many other factors, the costs of irrigation efficiency improvements are very likely to outweigh the economic benefits derived for many farmers. As a means for adding to the total water supply, the high costs and relatively small savings reduce the feasibility of this alternative.

However, as a method for improving water quality, especially reducing salinity, the costs are more acceptable.

One annual cost estimate for reducing salinity by improving irrigation efficiency is shown in Table 4-8. The cost to reduce the salinity one mg/1 at Imperial Dam by sprinkler irrigation ranges from $1.9 million to $4.1 million or $185 to $308 per ton of salt removed (1975 dollars). By comparison, canal lining would cost $140,000 to $370,000 annually per mg/1 reduction in salinity or $14 to $30 per ton of salt removed. Based on these figures, canal lining is a more cost effective salinity control measure than sprinkler irrigation. It should be kept in mind, however, that a desalting plant would cost about $30/ton per year; damages due to salt concentration are estimated at $20/ton or $230,000 per mg/1 (Gold and Calmon, 1980).

TABLE 4-8: Annual Cost of Salt Reduction by Improved Irrigation Efficiency in the Upper Colorado River Basin[a]

Method	Cost per Ton of Salt Removed (dollars/ton)	Cost per mg/1 Reduction at Imperial Dam (millions of dollars/mg/1)
On-Farm Improvements, Sprinkler Irrigation	185-308	1.9-4.1
Off-Farm Improvements, Canal Lining	14-30	.14-.37

Source: Utah State U., Utah Water Research Lab. 1975, pt. 2, pp. 260-62.

[a]1975 dollars

Estimates of the cost for reducing salinity in the Grand Valley of west central Colorado are given in Table 4-9. Each year approximately 660 to 760 thousand tons of salt reach the Colorado River as a result of saline subsurface irrigation return flows originating in the Grand Valley. The cost of using "best management practices" for reducing salinity one mg/l at Imperial Dam ranges from $0.34 million to $1.44 million, or $36 to $150 per ton of salt removed. "Best management practices" include: lateral lining, on-farm improvements (sprinkler irrigation, trickle irrigation), canal linings, and desalting. The selection of the alternatives is dependent on the amount of salt to be removed. For example, a reduction of 110,000 tpy of salt will require only lateral lining, while a reduction of 550,000 tpy requires all four alternatives. The cost for implementing each method is shown in Table 4-10.

TABLE 4-9:   Annual Costs for Reducing Salinity in the Grand Valley, Colorado

Reduction of Salt Load (tpy)	Salinity Reduction at Imperial Dam (mg/l)	Control Cost (millions of dollars)	Cost per Ton of Salt Removed (dollars/ton)	Cost per mg/l Reduction at Imperial Dam (millions of dollars/mg/l)
110,000	11.6	4	36	0.34
220,000	23.2	10	45	0.43
330,000	34.8	20	61	0.57
440,000	46.4	37	84	0.80
550,000	58.0	65	118	1.12
660,000	69.6	100	152	1.44

Source: Gold and Calmon 1980, p. 6.

TABLE 4-10:   Investment Costs for Improving
               Irrigation Efficiency

	Cost per Ton of Salt Removed
Method	(dollars/ton)
On-Farm Improvements	
Head ditch linings, gated pipes and/or automated cut-back furrow irrigation	90-100
Sprinkler irrigation and irrigation scheduling	120
Trickle irrigation and irrigation scheduling	200
Off-Farm Improvements	
Canal linings	170-630
Desalting and pump drainage	290

Source: Gold and Calmon 1980, p. 5.

　

Balanced against these increased costs are economic benefits of salinity reductions. The salinity problems in the CRB are already significant and are expected to increase in the future (Utah State U., Utah Water Research Lab. 1975). Damages to agriculture from salinity include reductions in crop yield, limitation of the types of crops which can be grown, and increased costs because of measures taken to avoid crop loss. As we have seen, alternative irrigation techniques and canal lining can substantially reduce salinity problems in the CRB. Therefore, agriculture would benefit economically from salinity reductions associated with improvements in irrigation. Unfortunately, there is no incentive for individual farmers to undertake these measures since the benefits of any salinity improvements would accrue to downstream water users. For this reason, salinity control efforts must be approached basin-wide (see chapters 3 and 6).

Implementation. Farming in most areas of the West,
especially the Southwest, is economically risky. Greater
capital and annual costs would make it even more risky.
In addition, the per-acre costs of converting to sprinkler irrigation and canal lining are very high compared
to the value of the agricultural land in many areas.
Thus, it is unlikely that many farmers could afford the
costs to convert to more efficient irrigation systems.
Subsidies from the U.S. Department of Agriculture (USDA)
and other federal agencies have been used in the past to
promote conservation and would be needed if this approach
were to be implemented on a large-scale basis. (See
Chapter 6 for a discussion of current efforts in the
context of salinity control.)

Further, in the current system, a high degree of
risk is associated with spending large amounts to improve
irrigation efficiency. This is because the appropriation
system may not allow a user to retain a right to conserved water. As discussed in Chapter 2, the doctrine
of prior appropriation in water allocation has often
worked as a disincentive to the efficient use of water
within the agricultural sector. Western water law based
on the doctrine of prior appropriation states that in
order for a water right holder to keep a right, the full
entitlement must be put to use or the unused portion may
be lost. This stipulation encourages the withdrawal of
all the water a user is entitled to even if the amount
exceeds what is required for consumptive use. In addition, some states require that the water be used only on
the land for which the right was initially acquired.
Therefore, the right holder cannot use the water elsewhere or attempt to transfer surplus water under the
penalty of losing the water permanently.

Improvements in irrigation efficiency may also have
adverse effects on wildlife habitat. Current irrigation
practices, particularly flood and furrow methods, have
created wildlife habitat in the farm irrigation ditches
and wetlands. The use of pipelines and reductions in
runoff would destroy much of this habitat. These potential consequences must be weighed against the benefit of
reducing salinity and leaving more water in the stream.

Conserving Water Through Crop Switching

Water can also be conserved by replacing water-
intensive crops, such as alfalfa and other forage crops,
with vegetables or wheats. For example, in Arizona,
alfalfa and crop pasture annually consume as high as 6.2
and 3.4 acre-feet per acre of irrigation water, respectively. In the same area, shallow and deep vegetables
consume less than 1.6 acre-feet per acre and wheat, bar-
ley, and oats, less than 2.2 acre-feet per acre. This

variation in consumption between crop types is similar throughout the study area (Gold and Calmon 1980, p. 132).

In general, the major crops grown in the study area are those which support extensive livestock production. These crops are forage crops, such as alfalfa and hay, which are used as winter feed for cattle and pigs. An agricultural profile of the study area is presented in Tables 4-11 and 4-12.

On farms with sales of $2,500 or more, a little more than 27 percent or 61,600,000 acres of the total farmland is cropland, with the remainder used predominantly for pasture or grazing. Approximately 61 percent of the total cropland is harvested. The total agricultural income from the study area was approximately $7.0 billion in 1974. Cropland income was $3.5 billion with livestock and livestock products contributing the remainder. The marketing of cattle and calves was by far the largest source of income within the livestock sector, accounting for about 79 percent of the total.

In general, grain production contributed the largest income within the cropland sector at 79 percent. This was followed by field seeds, hay, forage, and silage which accounted for 9 percent. These percentages refer only to those crops which were harvested and sold. Sixty-three percent of the farmland in the study area is used for livestock grazing and is not included in this category. Variations exist within the study area as shown by the data in Tables 4-11 and 4-12. For example, grain was the largest income producer in all

TABLE 4-11:   Farmland Use (in acres) in Seven Western States
(farms with $2,500 or more in sales)

	Total Land Area	Total Farmland	Total Cropland	Cropland Harvested	Cropland Used for Pasture or Grazing
Arizona	72,587,264	16,965,315	1,372,631	1,052,199	84,921
Colorado	66,411,072	33,739,546	10,207,361	5,892,090	1,113,667
Montana	93,157,952	56,731,193	15,237,374	8,375,116	1,040,944
New Mexico	77,703,168	38,017,791	1,900,964	926,931	370,707
North Dakota	44,334,726	40,792,094	28,997,608	19,158,983	2,194,445
Utah	52,540,672	8,725,508	1,668,708	1,024,528	360,818
Wyoming	62,212,224	30,376,101	2,555,196	1,654,395	503,652
Total	468,944,000	225,000,000	61,600,000	38,084,242	4,670,154

Source: U.S., Department of Commerce, Bur. of Census 1974.

TABLE 4-12: Value of Agricultural Products Sold[a]

	Total Crops	Grains	Cotton and Cotton Seed	Field Seeds, Hay, Forage, and Silage	Other Field Crops	Vegetables, Sweet Corn, and Melons	Fruits, Nuts, and Berries	Nursery and Greenhouse	Livestock	Total
Arizona										
dollar value	491								563	1,054
percent		15	42	11	5	15	11	2		
Colorado										
dollar value	681								1,281	1,962
percent		59	–	11	21	3	1	6		
Montana										
dollar value	549								476	1,025
percent		83	–	9	8	–	–	–		
New Mexico										
dollar value	154								354	508
percent		32	22	23	3	12	6	2		
North Dakota										
dollar value	1,437								362	1,799
percent		83	–	5	12	–	–	–		
Utah										
dollar value	91								237	328
percent		35	–	28	18	5	7	7		
Wyoming										
dollar value	122								237	359
percent		41	–	18	40	–	–	–		
Total										
dollar value	3,524								3,509	7,033
percent	50	64	6	9	12	3	2	1	49	

Source: Adapted from U.S., Dept. of Commerce, Bur. of Census 1974.

[a]Farms with sales above $2,500 in millions of dollars. Figures may not total because of round-off.

states except Arizona, where cotton was the largest income producer. North Dakota has by far the largest percentage of farmland used for crops, at 70 percent, compared to a 27 percent average for all other states.

As is the case in many of the western states, cropland provides a winter feed base for livestock production. Although dry cropland is an important source of feed hay and grain, irrigated cropland contributes the largest amount. North Dakota is the exception to this situation.

Several factors would be involved with crop switching as a comprehensive alternative to reduce water use in the agricultural sector. On the positive side, the economic return for vegetables is much greater than for forage crops; for example, in 1974 the value of vegetables, sweet corn and melon production in the study area averaged about $1,120 per acre as compared to the value of alfalfa production of $110 per acre. However, several other factors could threaten that economic return. First, it is unlikely that sufficient markets exist to support large-scale increases in vegetable production. Second, vegetables are more susceptible to weather damage, such as late freezes and floods, than are forage crops. The most important barrier to crop switching is that the major source of agricultural income in the study area is from the sale of livestock and their products, not crop production. Livestock enterprises account for 60 percent of the total market value of agricultural products sold. Alfalfa and other hays, corn and other grains are grown to support the cattle and dairy industries in the western study area. Many farmers have grown forage crops to support a livestock industry for decades. In the study area, vegetables are grown on less than 0.3 percent of the total harvested cropland. The farmers in this area are not likely to be willing to make radical shifts in crop type or cropping pattern. Further, the same institutional constraints which may not allow a farmer to have a legal right to water saved by improved irrigation efficiency could influence the decision to switch crops. For these reasons, although growing alternative crops may be used on a relatively small, localized basis to reduce water needs, large scale crop switching to transfer the economy from livestock farming to crop farming is not likely.

SUMMARY AND CONCLUSIONS

Table 4-13 summarizes and compares alternatives for conserving water in the energy and agricultural sectors.

Conservation in energy development appears the most promising in reducing conflicts over water availability. Changes in process design can substantially reduce consumptive water requirements and, depending on the price of water, water minimizing designs are economically attractive. Reductions in raw water requirements also reduce the volume of wastewater disposal. This represents an internal benefit to the energy company since investment and operating costs for wastewater treatment would be decreased. It is also an environmental benefit since chances of accidental spills are decreased. However, the total amount of wastes will not be reduced; rather, the concentration of wastes in the wastewater stream will be increased.

Since water for cooling is the largest single water requirement for electric power plants and synthetic fuel facilities, the choice of cooling technology is an integral part of water conservation. Water saving cooling technologies can substantially reduce consumptive water requirements. Wet/dry cooling of power plants can reduce consumptive water requirements by about 70 percent, while intermediate wet cooling of coal synfuel plants can reduce consumptive water requirements by about 20 to 30 percent. These water savings can increase economic costs. In the case of synfuel plants, however, the economic penalty for intermediate wet cooling is expected to be less than 0.5 percent. For power plants the economic penalty would typically be from 3 to 5 percent. In addition, with all dry cooling of power plants, plant efficiency and capacity could be lowered due to larger energy requirements to run the fans and due to higher condenser temperatures on hot summer days. For this reason, wet/dry cooling is generally much more practical than all dry cooling for power plants.

Although the economics are not always favorable, conservation techniques in energy facilities may be attractive for other reasons. In some water-short areas, water supplies may be physically or institutionally limited; thus water conserving technologies may be necessary. And, even if sufficient water is available, a facility with smaller water requirements will be less threatening to other water users. This factor is likely to become increasingly important in the siting process.

Regarding irrigation practices, considerable uncertainty exists about how much water can be "saved" because of questions about the amount and nature of irrecoverable losses in any given irrigation project. Nevertheless, it is generally agreed that the water "saving" potential is not as significant as the potential for improved water quality. In fact, current salinity control efforts in the CRB focus largely on improvements in irrigation efficiency. Also, by reducing the amount diverted, instream flow values will be

TABLE 4-13: Summary of Conservation Alternatives

Alternative	Current Status	Advantages	Disadvantages	Feasibility
Energy Process Design	No commercial coal synthetic fuel plants in existence; but water minimizing techniques included in many plant designs for the West	Economical reduction in water consumption in many cases Environmental benefits due to reduced diversions Reduction in wastewater disposal requirements	Potential increase in cost of final product	Included in many proposed facilities Possible need for additional incentives or regulations to enhance adoption
Cooling Techniques	Power plant with dry cooling operating at Wyodak, Wyoming	Substantial water savings capability Wet/dry cooling economical in some cases Environmental benefits due to reduced diversions	Cost of dry cooling 10 to 15 percent higher than wet cooling Lower plant capacity and possible lower plant efficiency with all-dry cooling	Adoption dependent on the price of water Possible need for additional incentives or regulations to enhance adoption Research underway to improve economics
Agriculture Improved Irrigation Efficiency	Local, state, and federal programs in progress now	Substantial reductions in quantity of water diverted	Uncertain, but probably small, reductions in water consumed	Probable need for additional subsidies to reduce costs to individual farmers

TABLE 4-13 (Continued)

Improved Irrigation Efficiency (continued)	Some salinity control projects currently in place include measures to improve Irrigation efficiencies	Substantial reduction in salinity	High initial capital investment	Programs aimed at improving irrigation efficiencies hampered by existing legal constraints and uncertainties over the rights to conserved water
		Increased crop production with sprinkler systems	Increased annual costs	
			Adverse impact on wildlife habitats dependent on return flow	
		Reduction in labor requirements	Effect of change in flow patterns on downstream users	
Crop Switching	Not generally practiced in the study area	Reduction in water diverted and consumed	Would require major shifts in current agricultural economy	Potential utilization on localized basis, but large-scale crop switching probably not feasible
	Water shortages forcing switch to less water-intensive crops or return to dry farming in some areas highly dependent on irrigation (for example, Texas and Oklahoma)	Higher economic return for vegetables than for forage crops	Limited market for vegetables	
		Would improve salinity levels	Possible alteration of support businesses for livestock production	

protected, but this will need to be balanced against the loss of wildlife habitat dependent on agricultural run-off. Although switching to less water-intensive crops could save large quantities of water, this is not a feasible region-wide policy option, given the current makeup of the agricultural economy.

Improvements in irrigation efficiency will face both economic and institutional obstacles. Generally, the economic costs far exceed the benefits to the individual farmer. If this option is to be implemented on a large scale, it will probably require additional federal subsidies. In addition, even if farmers conserve water, the "use or lose" doctrine means that water "saved" by the individual farmer cannot be retained.

Water conservation for energy resource development may be the easiest set of alternatives to implement. Conservation for energy resource development can save large percentages of water and has less uncertainty associated with its effectiveness and cost. It appears easier to implement new regulations on energy resource development than on agriculture, and energy industries are probably better able than farmers to afford the cost increases associated with water conservation efforts.

NOTES

[1]Once-through cooling and cooling ponds are two types of cooling which have been used in many other areas. However, they are generally not considered in the West because of their relatively large water requirements (see S&PP 1975, Chapter 12). In addition, once-through cooling for inland power plants is largely precluded by thermal pollution regulations under the Federal Water Pollution Control Act (FWPCA 1972).

[2]To convert to a per-thousand-gallon basis, note that $1 per thousand gallons is equivalent to $326 per AF.

[3]As one example, as mentioned in Chapter 2, the Intermountain Power Project is planning to purchase water rights for approximately $1,750 per AF. Assuming this right is exercised over 30 years, the purchase cost of water is roughly $58 per AFY.

[4]Larger evaporation losses occur with sprinklers than with furrow because the entire soil surface is watered as well as the surface of the plants. In addition, sprinkling with water containing appreciable amounts of dissolved solids (e.g., salt) could result in burn or death of plant leaves. Therefore, sprinkler irrigation would not necessarily be appropriate in areas with a high level of dissolved solids in the water. For further information see Doneen (1971).

5
Augmentation of Water Supply

INTRODUCTION

Augmentation of water supply is another set of options for mitigating water use conflicts among competing water users in the western states. Increasing water supply can be accomplished in conjunction with water conservation alternatives discussed in Chapter 4. Together these policy options respond to many of the problems and issues concerning water availability and quality identified in chapters 2 and 3.

Increasing water supplies can be approached in two ways. First, poor quality water can be developed for industrial needs, such as energy conversion facilities. An example of this approach is the use of saline water instead of fresh water in synthetic fuels plants. A second approach is to augment fresh water supplies in a river basin or watershed. Although water supplies could be developed specifically for energy use, simply making more water available in a basin could reduce conflicts over water availability. Alternatives include enhanced utilization of groundwater, weather modification, vegetation management, and surface water development projects. This chapter emphasizes saline water use, an option primarily applicable to the energy industry.

USE OF SALINE WATER BY ENERGY FACILITIES

Saline water, containing more than 1,500 to 2,000 ppm of total dissolved solids (TDS) is not suitable for many uses, particularly agricultural and environmental uses. Saline water is generally detrimental to irrigable land and some fishery resources, and the cost of treating saline water for these uses is uneconomical. However, energy developers can treat saline water and incur only modest increases in product costs.

143

Availability of Saline Water

Sources of saline water include both surface water
and groundwater. In the case of surface waters, tributaries of the Yellowstone River and many tributaries of
the Colorado River have saline flows. For example,
Table 5-1 identifies the major saline water sources in
the Colorado River Basin and their contribution to increased salinity at Imperial Dam as identified by the
Water and Power Resources Service. Consumptive use of
water from these sources has the potential not only for
supplying water for energy development but also for
reducing the salinity problems in the Colorado River
Basin identified in Chapter 4.

Sources of saline water in the western U.S. also include aquifers such as the Madison or the Dakota. Table
5-2 identifies the dissolved solids content from wells
in these two aquifers. As shown in Figure 5-1 saline
aquifers are present in much of the coal producing area
of the Upper Colorado River Basin (CRB). The suitabil-
ity of these aquifers as water sources for energy development depends on their ability to produce an adequate
supply. For example, although there are an estimated
255 million acre-feet (MMAF) of recoverable saline groundwater available in the San Juan River Basin (New Mexico,
Water Quality Control Comm. 1976, p. V-5), economic recovery in sufficient quantities may be difficult because
of low well yield (100 to 200 gallons per minute [gpm])
(Gray 1979). However, the Madison and Dakota aquifers
apparently have yields of 1,000 to 2,000 gpm.

Return flows from irrigation projects are another
source of moderately saline water (approximately 2,000
milligrams per liter [mg/l] TDS) which could be used by
energy development (Stone & Webster 1979b, p. 3). For
example, the Colorado River Salinity Control Forum is
considering industrial use of return flows from the
Navajo Indian Irrigation Project[1] (NIIP) as a part of
its salinity control program. For the proposed Sundesert Nuclear Plant in California, the San Diego Gas and
Electric Company signed an agreement with the Palo Verde
Irrigation District for the use of wastewater from the
outfall drain. To offset the increased consumptive use
of water, the company intended to restrict irrigation on
company-owned lands within the district (Calif., Public
Utilities Comm. 1976). Whether such offsets would be
necessary in all cases would depend on historical water
use patterns and state laws concerning downstream users'
rights to return flows. Because of the recent construction of the NIIP, no downstream users will depend on
return flows and, thus, no offsets will be necessary.

Municipal wastewater can also be used for energy facilities. However, the small size of western communities

TABLE 5-1: Saline Water Sources in the Colorado River
            Basin

Potential Source	Source Quantity (AFY)	Average TDS (mg/l)	Annual Salt Load (1,000 tons)	Effect at Imperial Dam (mg/l)
Big Sandy River	14,500	6,500	130	11
Meeker Dome	1,090	19,300	27	3
Glenwood-Dotsero	12,500	14,200	250	23
Grand Valley	43,500	3,300	190	20
Lower Gunnison	17,200	2,900	67	5
Paradox Valley	1,450	260,000	180	19
McElmo Creek	6,200	4,700	40	4
Uinta Basin	10,900	4,500	67	6
Price River	24,900	4,000	134	13
San Rafael	22,200	3,550	107	10
Crystal Geyser	150	14,000	3	1
Dirty Devil River	40,000	2,100	113	10
LaVerkin Springs	8,300	9,650	109	8
Lower Virgin	7,200	8,200	80	9
Las Vegas Wash	80,000	1,800	208	20
Palo Verde	187,000	1,700	420	39
Total	470,090		2,125	201

Source: Hinds 1980.

limits the usefulness of this source. A city with a
population of 70,000 has an average wastewater outfall
of about 7,000 gpm (over 11,000 acre-feet per year
[AFY]). Although this volume would sustain some energy
conversion units (see Chapter 2), there are few municipalities of that size in the study area. Naturally, if
a smaller city were located close to an energy facility,
the outfall from the sewage plant could provide a portion of the supply needed for operation.

Conjunctive use of saline and fresh water would increase the number of suitable saline water sources. However, designing a facility to handle saline water requires a large capital investment for the special treatment facilities. If the plant switched from high TDS
water to fresh water, those fixed charges would continue, leaving minimal economic savings to the plant.
Also, additional facilities would be required to handle
and condition the fresh water. Thus there are no significant incentives for conjunctive use in which feed water would be switched between fresh and saline sources.

TABLE 5-2: Saline Components of Water From Six Wells Which Might Serve Energy Converters

Components	Dakota Aquifer				Madison Aquifer		
	Two New Mexico Wells		An Arizona Well	A South Dakota Well	Two Montana Wells		A Wyoming Well
TDS, ppm	9,400	1,700	1,530	3,300	3,000	5,000	4,000
Calcium, ppm	30	12	5	18	340	350	320
Magnesium, ppm	90	3	2	7	51	53	120
Sodium, ppm	-	-	-	1,300	510	1,500	780
Bicarbonate, ppm	210	620	370	629	106	340	440
Sulfate, ppm	4,700	700	780	7	1,100	1,300	2,300
Chloride, ppm	1,400	400	27	1,600	660	2,000	160
Silica, ppm	-	11	10	10	42	55	13
Barium, ppb	-	-	-	-	-	100	-
Strontium, ppb	-	-	-	910	-	13,000	-

ppm = parts per million   ppb = parts per billion

Source: Selected data from the "Water Quality File" of the U.S. Geological Survey, Water Resources Division, Denver, Colorado.

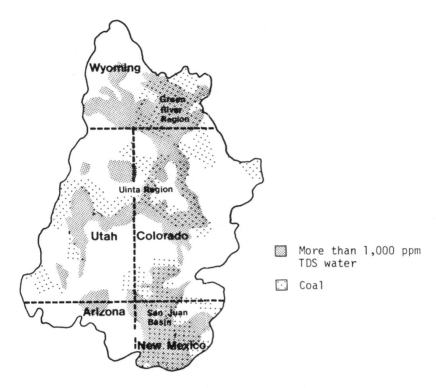

Figure 5-1: Coal and Shallow Highly Saline Groundwater, Upper Colorado River Basin

Source: Adapted from Radian 1977, pp. 39-40.

However, if joint, concurrent use of fresh water and saline water were employed, these disadvantages would be avoided. The two supply sources could be used in different parts of the plant. For example, if limited fresh water were available, it could be employed for processes requiring the cleanest water (e.g. boiler makeup water), and the balance of plant requirements could be supplied from saline water sources.

Costs of Saline Water Use

The economic costs for using saline water include supply, treatment, and disposal of brines. The cost of supplying water to energy facilities depends on the location of the water source, transportation distances, and, in the case of groundwater, depth and number of wells required. In 1979 dollars, cost estimates on supplying saline water from deep aquifers to energy facilities range from $34/AF (Missouri River Basin Commission 1978, p. 200) to $400/AF (Odasz 1979). Using the $400/AF estimate and not accounting for fresh water supply costs, the effect on total product cost for three energy facilities is shown in Table 5-3. The costs represent a small part of the final energy product costs.

To examine treatment costs, water usage in energy facilities can be placed into three categories: steam generation, process requirements, and cooling. Of these, boiler feed water has the most stringent quality requirements. In drum-type boilers operating at 2,000 pounds of pressure, makeup water must have less than 50 ppm TDS assuming 10 to 1 recycle ratio and maximum blowdown of 500 ppm TDS (Betz 1962). Resources Conservation

TABLE 5-3: Costs of Water Supply from Saline Aquifer

Process (capacity)	Total Water Usage AFY	Water Supply Costs ($/product)[a]
Gasification[b] (250 MMcfd)	14,000	0.06/Mcf
Liquefaction[c] (50,000 bbl/day)	10,000	0.22/bbl
Electric Power[d] (2,000 MWe)	19,000	0.00044/kWh

$/product = dollars per product unit
MMcfd = million cubic feet per day

Mcf = thousand cubic feet
bbl/day = barrel(s) per day
MWe = megawatt-electric
kWh = kilowatt-hour

[a]At $400 per AF; based on fresh water use values unadjusted for increase caused by change to saline sources.

[b]ANG North Dakota Project (U.S., DOI, BuRec, Upper Mo. Region 1977).

[c]Exxon EDS Process (Exxon 1977).

[d]Colstrip, Montana, Project (Westinghouse 1973).

Corporation (Radian 1977) has estimated the operating
and investment costs of a distillation-type process for
this quality of water to run about $5 per 1,000 gallons
(over $1,600 AF). Because the boiler water requirements
are small, total cost increases are low. As shown in
Table 5-4, boiler water treatment costs for a Lurgi-
based gasification process would increase gas costs by
$.01 per Mcf, a small fraction of the product costs.
Similarly, small cost increases are found for coal liquefaction and electrical generation.

Process water used to scrub particulates from gases
and to slurry lime for flue gas desulfurization, must be
similar in quality to cooling water. Thus, treatment
costs should be about the same (see Table 5-4). Costs
of treating moderately saline water (3,000 to 10,000 ppm)
for cooling tower use in a 1,000 megawatt (MW) power
plant can be as high as $.75 per 1,000 gallons (Radian
1977; Gold and Goldstein 1979). However, as shown in
Table 5-4, cost increase for the 2,000 MW Colstrip plant
would be $0.0005 per kWh (Balzhiser 1977, p. 41). This
would be a 1.25 percent increase for electricity generating costs of 4.0 cents per kWh. For coal gasification
and liquefaction, even smaller percentage increases in
product costs would result.

Use of saline water also increases waste handling
problems because the volume of brines produced from
boiler and cooling tower blowdown will increase. The
extent of the increase will depend on the composition of
the water and the treatment required before use. In addition, pretreatment may generate slurries and sludges
which also must be discarded. Generally, such brines
and slurries are sent to evaporative holding ponds where
solids and salts precipitate.

Table 5-5 presents one estimate of waste disposal
costs based on unit costs (Harza Engineering Company, as
cited in Mo. River Basin Comm. 1978, pp. 154-56). Because of uncertainty about the amount of extra brines
produced through saline water use, the original values
have been arbitrarily increased tenfold in Table 5-5.
However, even the resulting escalated costs are a small
fraction of product costs. An alternative technique for
disposal of brines is to inject them into a deep, very
saline aquifer (Barlow 1963, pp. 25-27). In both cases,
the large holding ponds or brine disposal wells will
increase the land requirements of the energy facility.
However, as noted by a study of the use by the proposed
Sundesert Nuclear Plant of wastewater from the Palo
Verde Outfall Drain, such requirements could be reduced
by 50 percent by operating cooling towers at 30 cycles
of concentration rather than the normal 15 cycles of
concentration (Stone & Webster 1979). While such operations would increase chemical treatment costs, the

TABLE 5-4: Added Costs for Treating Saline Waters[a]

Technology and Size	Boiler[b]		Process[c]		Cooling Tower[c]	
	Treated Water Demand (gpm)	Added Cost ($/product)	Treated Water Demand (gpm)	Added Cost ($/product)	Treated Water Demand (gpm)	Added Cost ($/product)
Gasification[d] (250 MMcfd)	500	0.01/Mcf	2,000	0.009/Mcf	6,300	0.03/Mcf
Liquefaction[e] (50,000 bbl/day)	1,000	0.1/bbl	2,700	0.06/bbl	1,300	0.03/bbl
Electric Generation[f] (2,000 MW)	200	0.00007/kWh	--	--	10,000	0.0005/kWh

[a] Escalated costs to 1979 at 12 percent per year.

[b] At $5 per 1,000 gallons, $1,600 per AF.

[c] At $0.75 per 1,000 gallons, $250 per AF.

[d] ANG North Dakota Project (U.S., DOI, BuRec, Upper Mo. Region 1977).

[e] Exxon EDS Process (Exxon 1977).

[f] Colstrip, Montana Project (Westinghouse 1973).

TABLE 5-5:   Costs of Disposal of Saline Waste

Technology and Size	Waste Handling Costs ($/product)[a]
Gasification[b] (250 MMcfd)	0.16/Mcf
Liquefaction[c] (50,000 bbl/day)	0.52/bbl
Electric Generation[d] (2,000 MW)	0.00014/kWh

[a]Harza Engineering Company (as cited in Mo. River Basin Comm. 1978, pp. 154-56); values in table have been increased tenfold.

[b]ANG North Dakota Project (U.S., DOI, BuRec, Upper Mo. Region 1978).

[c]Exxon EDS Process (Exxon 1977).

[d]Colstrip, Montana Project (Westinghouse 1973).

increase may be offset by a decrease in holding pond construction, operation, and decommissioning (Stone & Webster 1979b, p. 16).

Use of saline water in energy facilities may also increase corrosion problems (Fink 1963, p. 21). However, experiences of offshore petroleum production platforms using seawater for cooling indicate that this is not a serious problem. Although each facility will face somewhat different problems, it is estimated that cost increases will not be large even if titanium heat exchangers, used on offshore platforms, are necessary. Overall, because using saline water will probably require alloy heat exchangers, increased use of dry cooling may be attractive.

As summarized in Table 5-6, saline water can apparently be used by energy facilities at modest cost increases. However, in the case of saline aquifers, more information is needed to determine the full extent and cost of this potential source. Information needs include: (1) the characteristics of major saline aquifers; (2) the capacity of aquifers in probable energy development sites; (3) the composition of the water produced; and (4) impacts of aquifer drawdown.

Saline water use appears to have environmental benefits in limiting the withdrawal of fresh water supplies. However, brine disposal is a major environmental problem and additional land requirements are needed for evaporative holding ponds. Both withdrawal from and injection

TABLE 5-6: Estimated Costs of Utilizing Saline Water[a]

Category	Gasification[b] (250 MMcfd)	Liquefaction[c] (50,000 bbl/day)	Electric Generation[d] (2,000 MW)
Treatment			
Boiler[e]			
Need (gpm)	500	1,000	200
Cost ($)	0.01/Mcf	0.10/bbl	0.00007/kWh
Process			
Need (gpm)	2,000	2,700	N
Cost ($)	0.009/Mcf	0.06/bbl	N
Cooling Tower[f]			
Need (gpm)	6,300	1,300	10,000
Cost ($)	0.03/Mcf	0.03/bbl	0.0005/kWh
Disposal Costs ($)	0.16/Mcf	0.52/bbl	0.00014/kWh
Water supply costs ($)[g]	0.06/Mcf	0.22/bbl	0.00044/kWh
Total Costs			
$/product	0.27/Mcf	0.93/bbl	0.0011/kWh
Estimated prices ($)	4.50/Mcf	40.00/bbl	0.04/kWh
Cost as percent of price	6%	2%	3%

N = negligible

[a]From Tables 5-3, 5-4, and 5-5 of this report. No credit has been given for the normal treatment, supply, and disposal costs for facilities handling fresh water. Thus, the numbers here are worst case estimates.

[b]ANG North Dakota Project (U.S., DOI, BuRec, Upper Mo. Region 1978).

[c]Exxon EDS Process (Exxon 1977).

[d]Colstrip, Montana, Project (Westinghouse 1973).

[e]At $4 per 1,000 gallons ($1,200/AF).

[f]At $0.60 per 1,000 gallons ($200/AF).

[g]Based on water supply cost of $400/AF.

of brines into aquifers may also pose a hazard, as identified in Chapter 3. Nevertheless, because of the relative attractiveness of this alternative to agricultural, environmental, and municipal interests, increased saline water use by energy facilities is politically feasible in most western states.

Implementation Strategies for Saline Water Use

Although cost increases from using saline water appear to be less than 6 percent (Table 5-6), the private sector may need some incentive to pursue this option. In some cases, that incentive may be a lack of readily available fresh water. Such a situation provided the San Diego Gas and Electric Company the incentive for considering use of drain water from the Palo Verde Irrigation District (Cotton 1979). The Upper CRB is also a likely area for localized shortages which would encourage saline water use.

Energy developers may choose saline water to avoid costly delays in construction. At current construction inflation rates, costs for a one to two billion dollar facility increase one-half to one million dollars for each day of delay. Because the water rights transfer process in the West is slow, cumbersome, and costly, delays are almost certain. If political conflicts and lengthy water rights adjudication could be reduced by using saline water, economic savings through reduced construction cost could be realized.

Developers may choose saline water as a hedge against the uncertainty of the current water situation in the West. Since most industrial water rights will be junior to other uses, they are more susceptible to curtailment due to drought or federal/Indian water rights claims. If industry attempts to buy senior agricultural water rights, two problems may occur. First, there is a movement in some of the western states to use their water law to prevent transfers of agricultural water to industry (Trelease 1979). South Dakota, Wyoming, and Montana already have specific restrictions on transfers from agriculture to industry. Second, six of the eight states in the study area have preference clauses which allow water to be taken during shortages from less preferred uses (industry) for more preferred uses (municipal and irrigation), regardless of seniority of right. Since saline water is not useful for most other purposes, these provisions of the law probably will not restrict saline water use by energy developers.

States may also choose to encourage energy developers to use saline water as a part of an overall salinity control program, or to husband their diminishing supply of fresh water, or to protect irrigated agriculture. As

mentioned earlier, the Colorado River Basin Salinity
Control Forum is presently considering industrial use of
irrigation runoff in Grand Valley, Colorado, and the
NIIP in New Mexico to reduce salt loading in the Colorado River.

Several implementation strategies exist at the state
level to encourage energy developers to use saline water. The one which has the narrowest cost distribution
consists of allowing a regulatory body such as the state
corporation commission to adjust rates so that the increased costs are borne by the energy consumers. Such
action may be reasonable if a specific site could not be
developed to supply needed service without employing saline water. States served by the facility may determine
that higher rates are preferable to no service or to
restricted service from the plant.

A second strategy is for states to reduce the costs
for energy companies through some form of financial sub-
sidy--deferred income tax, lower property tax, or low
interest tax-free bonds to finance the commercial unit.
Although this option places the burden on the state population where the facility is located, the increased
revenue from the plant, the employees, and the supporting infrastructure may justify it.

A third strategy, which distributes increased costs
nationally, is federal action to provide incentives to
energy developers to use saline water. For example, the
Colorado River Basin Salinity Control Forum has recommended that the U.S. Water and Power Resources Service
consider encouraging the industrial use of saline water
where cost effective as a joint private-government salinity control measure (CRB Salinity Control Forum 1980).[3]
Related specific options include special tax credits for
equipment specifically needed to treat saline water,
special loans at reduced interest for the same equipment,
or even a direct subsidy to make energy from plants using
high TDS water more competitive in the market. For energy facilities which would serve primarily national energy needs, such as those built primarily to export
energy beyond the eight-state region, this type of cost-
sharing may be justified. In addition, Region VIII EPA
policy strongly encourages energy developers to utilize
saline water supplies (U.S., EPA, Region VIII 1980).

GROUNDWATER USE AND STORAGE

Water in both shallow alluvial aquifers and deep
aquifers is frequently suggested as a water supply alternative. Considerable quantities of such groundwater
exist in the eight-state area: for example, an estimated 50 to 115 MMAF is contained in Upper CRB aquifers

at a depth of less than 100 feet, of which about 15 percent is recoverable (Price and Arnow 1974). Agriculture and municipalities already use this source in many areas of the eight-state region. Some energy projects, such as the Black Mesa (Arizona to Nevada) and proposed Energy Transportation Systems, Inc. (Wyoming to Louisiana) coal slurry pipelines, use or plan to use water from deep aquifers.

While groundwater could increase existing supplies substantially in the eight-state study area, the interdependence between streamflow and alluvial aquifers may limit withdrawals. Not only would depletion of surface flow have possible ecological impacts, but the water rights of other users of the stream might be impaired. Further, if deep aquifers are mined continuously, future water users will pay the costs of present use. In most of the eight states, rights to groundwater are appropriated in the same way as surface water rights. An exception is Arizona where groundwater is treated as a private resource, conveyed with the land. Thus, in Arizona, groundwaters recharged from the surface become landowners' property and are outside government control.

Conjunctive use of groundwater and surface waters may provide a more uniform year-round water supply without the need for surface reservoir construction. That feature is particularly attractive to energy conversion plants since their demands are not seasonal, except for power plants with wet/dry cooling. It would decrease land needed for reservoirs, be less damaging environmentally, and require less capital investment. Perennial streams which are underlain by permeable alluvial aquifers may provide the opportunity for using groundwater in conjunction with surface sources (Price and Arnow 1974, p. C-26). Several suitable aquifers exist in the Upper CRB which may be useful in coal and oil shale operations. However, because the requirements for successful conjunctive use without damaging streamflows are restrictive, this option for augmenting surface water supplies during seasonal low flows will be determined on a site-specific basis.

Another approach to water augmentation is underground storage. An estimated 2 MMAFY evaporate from major reservoirs and lakes of the CRB (U.S., DOI, BuRec, and USDA, SCS 1977, Vol. I, pp. 11-30). However, augmentation by groundwater storage is geographically limited by the location and accessibility of large aquifers. At suitable sites, groundwater storage and use are often less expensive than transporting surface water over long distances, although considerable investments are necessary for pumping systems to store and/or retrieve the water. Estimates of groundwater mining costs vary widely; $34 to $46 per AF in the Missouri River Basin (MRB) (Missouri River Basin Commission 1978,

p. 200), $45 to $70 from the Madison aquifer in Montana (Montana DNRC, Water Resources Division 1976), and $400 for Energy Transportation Systems (Odasz 1979) for deep wells to the Madison aquifer. Costs for storage can be twice as high depending on how water is introduced into the aquifer (U.S., Army Corps of Engineers 1978). If aquifers are shallow, cisterns or percolation ponds can be used to reduce recharge costs.

In addition to reducing evaporation losses, using underground storage has environmental benefits. If wells are properly cased to prevent crossflow between aquifers, the system is essentially benign. It has the advantage of requiring much less land than surface storage and therefore has less impact on the local ecology.

WEATHER MODIFICATION

In the eight-state study area, weather modification (cloud seeding) is being studied as a way to increase the snowpack in mountains above 8,000 feet. Contrary to popular belief, weather modification is not a technique which can be used to end severe droughts. It may, however, increase the precipitation from natural cloud systems by initiating or extending the duration of snow or rainfall.

So far, the small pilot experiments (NAS/NRC 1973, p. 80) have not provided a firm base on which to estimate the enhancement resulting from seeding. Weisbecker (1974, p. 17) estimated that seeding could increase the snowpack by 20 to 25 percent, but more recent figures (Silverman 1980) are in the 10 to 15 percent range. Even that lower figure still gives increased snowpack runoff of about one and one-half MMAF in the Upper CRB. In 1971, cost figures for augmented snowpack runoff of 2.3 MMAF were about $2.4 per AF ($6, 1979 dollars) (Weisbecker 1974, p. 29); in 1975 another study estimated costs of about $5 per AF ($7.5, 1979 dollars) (U.S., DOI, BuRec 1975, p. 442). As these figures indicate, even with the large uncertainty in the magnitude of the added runoff, weather modification appears economically attractive. However, these costs do not include costs which may be required to deal with possible adverse impacts.

In order to remove uncertainties in the physical, socioeconomic, and ecological aspects of snowpack augmentation, a program has been developed by the Water and Power Resources Services (WPRS) to obtain needed data (Silverman 1980). WPRS expects to be able to forecast behavior from cloud seeding in five subbasins in the Upper CRB and to observe some of the longer range ecological effects on the flora and fauna of the region. One of the benefits from the increased snowpack runoff may be decreased salinity in the river main stem.

Because water users benefitting from weather modification may not bear the costs of the program, considerable controversy has developed. Among the potential problems are: (1) greater danger from avalanches; (2) higher flood levels during spring melting; (3) higher snow removal costs; (4) greater feeding costs for ranch and wild animals; (5) forced changes in pasturing plans through prolonged snow cover; and (6) reduced precipitation in downwind areas. If the WPRS program does demonstrate that the effects of cloud seeding meet expectations, it will require new regulations and adequate enforcement to minimize these risks and to provide compensation for adversely affected parties.

Another problem posed by weather modification involves the rights to the increased water supply. While the courts have acknowledged the rights of a developer of "new" water to use it (Fischer 1975, p. 643), the developer has the burden of proof that he did create "new" water. To that end, experiments and related modeling efforts would be required to validate claims of rights to water resulting from weather modification. Although Arizona legislation goes along with the concept of rights to "new" water, both Utah and North Dakota laws treat augmented precipitation the same as natural sources and thus deny benefits to the developer of the additional water (Davis 1979, p. 848).

Further restrictions on the rights to water from increased snowpack arise from the compacts and treaties covering regional and basin water resources. The wording in those agreements, written before weather modification was envisioned, is such that "new" water may not be outside the restrictions of those covenants (Fisher 1975, p. 656). For example, based on the terms of the Colorado River Compact (1922), only about one-half of the runoff from augmented snowpack in Colorado could be used within the state; the remainder by agreement would go to the benefit of Utah and Arizona. Since Weisbecker has estimated augmented snowpack could produce enough snow-melt runoff to provide all the water supply required by treaty for Mexico, "the potential advantages which could accrue through a cooperative effort between the states and the United States are enormous" (Fischer 1975, p. 556).

VEGETATION MANAGEMENT

Vegetation management involves removal of natural vegetation to increase runoff or reduce water loss by evapotranspiration. In uplands, this requires removal of trees and shrubs. Tree removal can increase runoff by patterned clear-cutting to provide snow traps and by simply reducing tree density. Upland vegetation management has widespread potential throughout the eight-state study area, although it is estimated to be most effective in the forest lands of Montana and Wyoming (U.S., National Water Comm. 1973, p. 357). In lowlands, phreatophytes (deep-rooted trees such as chaparral which consume large volumes of water and which generally have low economic value) are removed from streams, stream beds and along canals, irrigation ditches, and other man-made water courses to decrease water loss. The removed vegetation can be replaced with other forms of vegetation, such as grasses, which consume less water.

The potential for vegetation management to increase runoff is based on the quantities of water lost to evaporation and transpiration by plants—approximately 90 percent of the annual precipitation in the CRB. At a maximum, according to Hibbert,

> ...flow of the Colorado River could be increased as much as 4 million acre feet [per year] in the Upper Basin and 2 million acre feet [per year] in the Lower Basin if all forest and brush lands were managed solely to increase water yield. However, these are unrealistic goals when other forest resource, economic, social, and environmental concerns are considered [1979, p. 1].

If a limited quantity of water would be desired to meet the needs of a specific project, Hibbert (1979, p. 18) has estimated that 42,000 AFY could be added by managing 250,000 acres of subalpine forests in the Upper CRB or 210,000 acres of chaparral in the Lower CRB.

Direct costs of upland vegetation management have been estimated at about $2 (1979 dollars) per AF for commercial forests in Montana, Wyoming, Colorado, Utah, New Mexico, and Arizona (U.S., National Water Comm. 1973, p. 357) and up to $100 (1979 dollars) per AF in chaparral areas. Phreatophyte removal costs have been estimated at $28 per AF (1979 dollars) for areas in Colorado, Utah, Arizona, and New Mexico (U.S., National Water Comm. 1973, p. 357). Upland vegetation management would be a mid- to long-term augmentation option; Hibbert (1979, p. 19) suggests that extensive water yield improvements from subalpine forest management in the Upper Basin would require several decades to become fully operational. However, conversion of chaparral to grass in the Lower Basin could yield additional water within a few years; pilot programs are currently underway (Hibbert 1979, p. 19; Brown et al. 1974).

The largest costs and risks for vegetation management appear to be environmental. The extent of possible
impacts is not clear, but current pilot programs should
reduce the uncertainty. Replacement of trees and shrubs
by grasses in upstream riparian areas could seriously
threaten the most productive wildlife habitat in the
Southwest, as well as impair recreational and scenic
values. Further, such riparian vegetation management
could increase stream temperatures causing harm to the
fish population (U.S., National Water Comm. 1973, pp.
354-58). Upland vegetation management in forest areas,
particularly clearcutting of forests which would contribute the largest runoff, could also threaten wildlife
habitats and degrade scenic and aesthetic values. Also,
the removal of vegetation will probably increase stream
velocities and erosion, causing increased silting downstream and possible flooding. Permanent control of
deep-rooted chaparral shrubs which sprout vigorously
from their crowns is difficult without extensive use of
herbicides. Thus, all of these options could be expected to generate environmental opposition.
Legal problems over rights to water from vegetation
management are potentially as difficult as those resulting from weather modification. Riparian vegetation
control would yield water increases which probably would
be considered "salvaged" water. In general, the right
to "salvaged" water does not accrue to the developer but
is subject to the priorities of the appropriation system. Upland vegetation management, such as snow fences,
designed to hold snow which might otherwise blow away
and evaporate could produce water which might be classified as either "new" or "salvaged."

SURFACE WATER STORAGE, DIVERSION, AND TRANSFER

The Colorado River now has reservoirs and water projects with a capacity of over 60 MMAF. They play a critical role in maintaining stream flows and supporting
irrigated agriculture. The potential for additional
projects is very uncertain because of a variety of political, environmental, and economic questions. And, even
if additional projects were built, considerable uncertainty exists regarding the effect on water availability
problems and issues. Additional storage capacity does
not increase the water supply since it serves only as an
averaging mechanism--collecting water during high flow
periods for release during low flow times. The ability
to control the timing of water availability through surface storage is essential to all water users.

However, several questions emerge regarding the future use of this alternative. First, there is a penalty to using impoundments; water is lost due to evaporation. In the CRB, about 14 percent or 2 MMAFY, of the average annual flow is lost to evaporation (U.S., DOI, Water for Energy Management Team 1974). Second, in the CRB so many impoundments already exist that few acceptable locations may remain (U.S., National Water Comm. 1973, pp. 319-33). Several potential storage sites do exist on the main stem of the Yellowstone River (see Table 5-7); for example, the Allenspur site, considered the best remaining site for a dam in Montana, has a potential usable storage capacity of 1.7 MMAF (Boris and Krutilla 1980, p. 71).

Transfers of water within basins (intrabasin) have been suggested as a way to increase water supplies in localized areas, particularly energy-rich areas of the CRB. This option may become more viable as the need for domestic energy development increases. Intrabasin transfers have been considered in southern and central Utah and western Colorado (U.S., DOI, BuRec 1975, pp. 143-49; NPC 1972, pp. 245-48).

Interbasin transfers are technologically feasible and potentially large sources of water, particularly if water from the Columbia or Upper MRB were diverted to the Upper Colorado. However, several major obstacles exist to this option. First, political opposition has been strong enough to prevent this alternative for the past twelve years through the Colorado River Basin Project Act (1968, renewed in 1978). This act prohibits the Department of Interior (DOI) from even carrying out any "reconnaissance" (investigative) studies of interbasin transfers from any other basin to the CRB. Further, transfers from the Yellowstone would require agreement among signatories (Wyoming, Montana, and North Dakota) to the Yellowstone River Compact (1950).

Second, interbasin transfers are expensive. One study that examined transfer of 2.4 MMAFY from the Snake River to the Colorado estimated costs at over $5 billion (1979 dollars)[2] (U.S., National Water Comm. 1973, p. 317). Another possibility, transferring 15 MMAFY from the Columbia main stem to the Colorado, had cost estimates of over $40 billion (1979 dollars) (U.S., National Water Comm. 1973, p. 317). Including operation costs, these two transfer projects would have provided water at costs between $2,000 and $4,000 per AF (1979 dollars).

More precise estimates of costs for specific projects depend on interest rates, flow rate through pipelines, and on locally variable construction and operation costs. According to Gold and Goldstein (1979), pipelines carrying 30 AF per day (about 10,000 AFY) will have operating costs ranging from $3 to $7 per AF per mile. Larger scale projects, transporting about 1,000

TABLE 5-7: Selected Potential Onstream and Offstream
Storage Sites, Yellowstone River Basin,
Montana

Name	Stream and Location	Drainage Area (square miles)
Upper Subbasin		
Allenspur	Yellowstone above Livingston	3,620
Wanigan	Yellowstone south of Livingston	3,114
Absaroka	Yellowstone below Livingston	4,778
Billings Area Subbasin		
Canyon Creek	Offstream near Billings	106
Buffalo Creek	Offstream near Custer[a]	229
Bighorn Subbasin		
Custer	Bighorn River near mouth	22,410
Mid-Yellowstone Subbasin		
Lissa	Yellowstone near Bighorn	37,315
Tullock Creek	Offstream near Bighorn River	458
Cedar Ridge	Offstream near Hysham[a]	37
Sweeney Creek	Offstream near Forsyth	99
Moon Creek	Offset near Miles City	66
Kinsey Area Subbasin		
Sunday Creek	Offstream near Miles City[a]	716
Lignite Creek	Offstream near Miles City	8
Tongue Subbasin		
New Tongue	Tongue River at 4-Mile Creek[b]	1,770
Stateline	Tongue River at Wyoming-Montana Line	1,115
Powder Subbasin		
Moorhead	Powder River near Moorhead	8,030

Source: Adapted from Northern Great Plains Resources Program
1974.

[a]BuRec (The U.S. Water and Power Resources Service) was granted a
reservation of water for this project from the Montana Board of
Natural Resources and Conservation.

[b]The Montana Department of Natural Resources and Conservation was
granted a reservation of water for their location by the Board of
Natural Resources and Conservation.

AF per day, cost $0.60 to $1.10 per AF per mile (WPA 1978). The Bureau of Reclamation (BuRec) studied the feasibility of diverting water from the Green River in the Upper Colorado to Gillette, Wyoming, 225 miles away via the North Platte River and pipelines. Assuming existing reservoirs are used, capital costs for pipes and pumping stations would be about $1 billion (1979 dollars) and operating costs about $6 million per year (1979 dollars). Total project costs would be about $300 (1979 dollars) per AF (U.S., DOI, BuRec 1972).

A third major concern with this option is environmental. Surface water diversions and storage projects have been criticized because of damage to scenic and wilderness areas and decreased stream flows in areas of the source of the diversion. In addition, in many stream segments, impoundments and diversions are precluded under provisions established by the Wild and Scenic Rivers Act (1968). While the Allenspur site is considered to be the best site for a dam in Montana, it "lies in the middle of a 95-mile blue ribbon reach of the Yellowstone River, renowned for its aesthetic value and its native trout fishery" (Boris and Krutilla 1980, p. 71). When an impoundment was suggested for that site, opposition was so great that the Montana legislature passed a joint resolution declaring that construction of the dam would be contrary to state goals and objectives. However, surface impoundments do provide a direct environmental benefit in low flow years by helping to maintain minimum flows, and also provide recreational opportunities for the public.

Questions have also emerged regarding how equitably this alternative will distribute costs, risks, and benefits. It is difficult, if not impossible, to calculate in economic terms the range of values involved in transferring water from one basin or subbasin to another. The values of water for economic growth, hydroelectric production, recreation uses, commercial traffic, supporting aquatic life, and providing scenic and aesthetic amenities may need to be compared to the values of transporting water to support energy resource development (U.S., National Water Comm. 1973, pp. 319-31). Even if these values could be adequately expressed in economic terms, considerable uncertainty exists regarding long-term projections of costs, including construction costs and cost recovery of these projects.

Some general concerns regarding who will pay and who will benefit can be addressed. Costs for transferring and storing water will ultimately be paid by the taxpayer since the capital and operating costs will probably necessitate federal authorizations. This would be equitable if water is transferred for energy resource development in order to meet national energy goals.

Since areas receiving the water would gain other benefits, for example, those associated with economic
growth, they should expect also to pay increased costs.
On the other hand, the basin from which the water is
transferred would generally be the loser--hence the
strong political opposition that usually exists to such
schemes. If the ultimate users of water are asked to
pay the costs of water augmentation, energy industries
could probably afford most of these water prices, while
agricultural interests would generally be unable to
afford them.

SUMMARY AND CONCLUSIONS

This chapter has reviewed a variety of technological
options for adding to the available supply of water.
Some of these alternatives are intended to provide new
sources for energy development projects. Others are
intended to increase the overall availability of water
in a basin. In both cases, the objective is to reduce
the potential conflicts involving water use and water
quality in the West. Table 5-8 summarizes the advantages and disadvantages of these augmentation alternatives.

When the criteria of (1) magnitude of supply, (2)
environmental impact, and (3) implementation feasibility
are considered, use of saline ground and surface waters
appears attractive. The effectiveness of this option is
site-specific because both the energy resource and the
saline water must correspond as, for example, in areas
where coal is found close to the Madison aquifer. Where
it is applicable, it can provide new water supplies to
energy conversion facilities at a maximum increased product cost of 2 to 6 percent. Although there may be some
problems with increased waste disposal, this option
could produce major environmental and economic benefits
as part of salinity control programs.

Groundwater use, either separately or conjunctively
with surface water, can provide a short-term, localized
benefit, particularly for energy facilities. However,
in some areas, groundwater is already being used by
agriculture and municipalities at rates beyond annual
replenishment; therefore, these sources would probably
be unavailable to meet energy needs. There are deep
aquifers underlying shale and coal resources which could
be developed rapidly for energy conversion facilities.
Further study would be needed on a site-by-site basis to
provide information on the ability of these aquifers to
sustain withdrawal rates during the 20- to 30-year life
of the commercial units. While mining beyond yearly
replenishment rates is a serious issue, it may be

TABLE 5-8: Summary Evaluation of Augmentation Alternatives

Alternatives	Current Status	Advantages	Disadvantages	Feasibility
Use of saline water by energy developers	Technically feasible Not currently being used by energy developers	Potential for adding thousands of AF to existing supplies Less competition for saline water than fresh water Potential environmental benefit if saline surface flows are reduced, particularly in the CRB Potential available within short term No legal changes required	Probable higher costs, but maximum penalty of 2 to 6 percent Increased waste disposal problems Limited knowledge of saline aquifers and the environmental impacts of their depletion	No legal impediments to natural saline water use; possible impairment of downstream users' rights if irrigation runoff used No existing state programs to encourage use Benefit to salinity control programs through use of irrigation runoff and other saline surface waters
Groundwater uses	Widespread use throughout the West	Could provide millions of AFY on a short-term basis Costs ranging from $45-$70 per AF Elimination of need for surface storage through conjunctive use of alluvial aquifer during low stream flow; protection of instream values	Long-term potential for depleting available supply and limiting surface water recharge if groundwater is mined Constrained by the location of suitable aquifers Costs potentially higher than surface water sources	Infrastructures in place to handle present usage of sources Risks borne by future generations if groundwater is mined Impeded in some states by existing groundwater laws which put groundwater outside of government control (Arizona)
Groundwater storage	Not currently being used on a large scale in the U.S.; some groundwater recharge programs used in California	Reduction of evaporative losses compared to surface storage No apparent adverse impacts on the environment Storage of water during high flow periods for later use	Limited by the location and accessibility of large aquifers Costs generally higher than for surface storage ($150 per AF if aquifer is deep) No recreational benefits	Possible need for federal subsidies to cover large investments Lack of political rewards for relatively invisible storage project which provides no recreational opportunities

TABLE 5-8: (Continued)

Weather modification	Being tried in some areas of U.S., but actual effects uncertain	Potential for adding large supplies, particularly in Rocky Mountain area Estimated costs about $5 per AF Great flexibility in terms of capital costs and reversibility of decision	Uncertainty about potential success; 9-year research program required to obtain statistically significant data Long-term physical, socio-economic, and ecological impacts uncertain Does not add to supply during droughts; only extends precipitation periods	Risks borne by future generations Probable opposition from environmental interests Benefits for "new" water developers overshadowed by present compacts and treaties; state and federal cooperation required
Vegetation management Runoff modification Removal of nonproductive vegetation	Not used on a large scale	Potential for saving millions of AF currently lost to evaporation and transpiration by plants Costs roughly $2 per AF for commercial forests in Montana, Wyoming, Colorado, Utah, New Mexico, and Arizona, and up to $100 per AF in chaparral areas Small capital requirements	May damage scenic and wilderness areas, including wildlife habitats If chemically controlled, potential damage to surface water quality Limited to 16 percent of CRB since much of basin has sparse vegetation Potential for increasing flooding and erosion problems with severe silting; long time (decades) to become operational	Probably affordable by agricultural interests Probable opposition from environmental interests concerned with wildlife habitats Benefits denied to developer because of legal complications based on rights to "salvaged" water

(continued)

TABLE 5-8: (Continued)

Alternatives	Current Status	Advantages	Disadvantages	Feasibility
Surface water diversion, transfer, and storage	Traditional approach to water resources management	Hundreds of thousands of AFY possible through intrabasin transfers (particularly in the Upper MRB) Intrabasin transfer costs potentially feasible for energy use Millions of additional AF possible over the long term through interbasin transfers	Restricted suitability of intrabasin transfers in CRB Expense of interbasin transfers, e.g., $2,000 to $4,000 per AF for transfers from Columbia and Snake rivers to CRB Potential damage to scenic and wilderness areas Serious equity questions with intra- and interbasin transfers Necessity of long-term and irreversible allocation of capital	Strong political opposition, e.g., DOI currently prohibited from studying projects to transfer water from the Columbia River to the Colorado; consent from Wyoming, Montana, and South Dakota required for transfers from the Yellowstone Some transfers limited by Wild and Scenic Rivers Act particularly in the Upper CRB Federal subsidies required because of billions of dollars in capital costs

acceptable in particular areas until other sources can be found. Regulatory facilities and appropriation procedures are already in place to handle groundwater allocations to industry.

Both weather modification (snowpack enhancement) and vegetation management can increase water supplies by one to two MMAFY in the CRB. Although economic costs for these options appear to be very low, there may be secondary costs adversely affecting others in the watershed that would have to be carefully evaluated. Further, a comprehensive analysis of the ecological and environmental impacts of these alternatives would be needed before any program could be developed to implement these augmentation processes. The time span required to achieve fully operational programs reduces the feasibility of these options. Serious legal questions concerning the rights to the increased water supply exist for both options. New legislation may be needed to assure that the benefits from weather modification or vegetation management accrue to those who produce the additional water supply.

Interbasin and intrabasin transfers and increased storage can increase usable water supplies in the eight-state region. These long-term options are costly both in environmental damage and in capital requirements. These alternatives continue to face substantial political opposition, particularly legislation which prohibits DOI from studying water transfer to the CRB. In addition prohibitions on development may restrict water transfers, such as in the Yellowstone River, where instream requirements limit the construction of diversions and impoundments.

The consistent conclusion regarding the alternatives discussed in this chapter is that augmentation alternatives are very costly ways to deal with the complex problems and issues affecting western water resources. Substantial quantities of water can be added by these alternatives, but when considering economic costs, environmental damage, and long-term ecological risks, several of these choices are likely to worsen rather than improve conflicts over the appropriate use of water. This conclusion represents a substantial change from past approaches to western water problems which largely emphasized technological fixes to get water where it is needed. Although such approaches have proven valuable in the past, they appear to be very limited as solutions to future problems. The exception to this conclusion is the use of saline water by energy facilities, which in many areas can provide important water availability and quality benefits without large costs or risks. To a lesser extent, groundwater use can be a feasible alternative for augmenting water supplies on a short-term, localized basis.

168

NOTES

[1]The Navajo's water right in the Navajo Indian Irrigation Project (NIIP) is for consumptive use only. Under this plan return flows belong to the state of New Mexico.

[2]Costs escalated at 12 percent per year compounded from year of original publication.

[3]The name of the Bureau of Reclamation was changed to the Water and Power Resources Service in late 1979. On May 20, 1981, Secretary of the Interior James Watt restored the original name.

6
Water Quality Protection

INTRODUCTION

Chapter 3 described the ways in which water quality can be adversely affected by energy development. Energy extraction and conversion activities can disrupt and contaminate aquifers and generate large amounts of wastes which must be disposed of to minimize leakage or leaching into surface waters and groundwaters. These potential water quality problems are difficult to quantify and the current level of knowledge about the risks and control measures is inadequate. Problems arise not because individual energy development activities will discharge large amounts of pollution directly into surface waters but rather because energy facilities are likely to be widespread in the region and because of the variety of chemical contaminants which will need to be isolated from the environment. Salinity is also an important water quality problem. Although the major sources of salinity are natural salt flows and agricultural runoff, energy development can contribute to the problem by increasing salt loadings and by consuming water which can increase salinity concentrations downstream. Another water quality concern, inadequate sewage treatment, is created by the rapid population growth in small western communities near energy development activities.

Throughout this book, it is emphasized that water use and water quality are intimately related. Thus, policies aimed primarily at dealing with the availability of water can also have important implications for its quality. For example, improving the efficiency of agricultural water use can provide both water availability and water quality benefits (see Chapter 4). The purpose of this chapter is to examine four policy options which are primarily intended to deal with water quality issues associated with energy development. These are:

- Improve planning and information for energy development including predevelopment baseline data, technological controls, and research;

- Allow temporary sewage treatment measures for communities affected by energy development;

- Construct salinity control projects; and

- Institute a "salinity offset policy" for energy development.

Since the long-term risks to water quality from energy development activities are so uncertain, the first policy option listed above is primarily aimed at improving our technical understanding. Increased monitoring and research activities will allow better technological controls to be devised for both resource extraction and waste disposal. The second policy option addresses problems of municipal sewage treatment in small communities that are affected by energy development projects. The goal of this policy alternative is to allow temporary treatment measures (sewage lagoons) until more conventional secondary and tertiary treatment facilities can be constructed. The third policy option is the primary strategy being followed to deal with salinity problems in the Colorado River Basin (CRB). We have included it to provide a point of comparison for other options which have an effect on salinity, including water conservation in energy facilities and improved agricultural water use, both discussed in Chapter 4. The fourth policy option, the "salinity offset policy" is a broad, institutional approach for dealing with salinity problems but is discussed here only as it applies to energy development. It is modeled after the air emissions offset policy used by the Environmental Protection Agency (EPA) in areas where pollution levels are above national standards. The basic concept of a "salinity offset policy" is that energy development projects having a measurable effect on salinity would be allowed only after some "offsetting" measure is found to prevent increased salinity levels.

IMPROVED WATER QUALITY CONTROLS

Improving existing water quality controls is intended to minimize threats to surface and groundwater quality and to increase the information base so that a better understanding of the environmental risks can be attained. Three specific ways to improve the regulatory system are discussed:

(1) Predevelopment monitoring of groundwater resources and quality;

(2) Water quality control plans for each energy facility; and,

(3) Monitoring and research of the water quality effects of energy development and control techniques. This option would build upon current U.S. Geological Survey programs.

Monitoring and research are especially important due to the high degree of uncertainty associated with groundwater quality. The degree of protection required, the effectiveness of control techniques, and the long-term assurance of high quality groundwater supplies are unclear. Monitoring would improve knowledge about geohydrologic behavior and would help to verify the effectiveness of control technologies. Data developed in a comprehensive monitoring program would be useful to the research community and, as more effective control techniques are developed, to the policy community.

The following section will describe and evaluate the three elements of this policy option. Although each element might be considered as a separate policy choice, together they constitute a comprehensive water quality control policy. Some of the elements are included in existing federal and state legislation. However, their effectiveness varies considerably between surface and groundwater, among the six energy resources, and among the eight states.

Predevelopment Monitoring

Considerable uncertainty exists about the quality of both surface water and groundwater prior to energy development. For example, in some locations in the Grants Mineral Belt in New Mexico, levels of selenium and radioactive isotopes found in groundwater are significantly higher than in other areas of the Southwest (U.S., EPA n.d.). However, it is not clear to what extent mining activities are responsible for these elevated levels. The large degree of local variability of groundwater quality prior to disturbance is a problem in establishing regulatory guidelines. Predevelopment groundwater monitoring can provide essential baseline

---

MONITORING INADEQUATE

In New Mexico, for example, more than 30 million gallons of water per day are discharged from uranium mines, affecting mine water and groundwater and surface waters below the discharge. The New Mexico Environmental Improvement Division has documented elevated levels of radiation and trace elements in groundwater adjacent to these activities, and has indicated that "adequate monitoring of the movement and the seepage and of the injected wastes is not underway...." This has led staff members of the agency to indicate the "imminent necessity for adequate monitoring and surveillance and the implementation of adequate regulatory controls to control unchecked degradation of the environment resulting from uranium mining and milling activities."

--U.S., EPA n.d.

---

data for determining appropriate controls, for establishing standards for permits and monitoring, and for determining liability for subsequent changes in groundwater quality.

Requirements for predevelopment water quality monitoring in both surface waters and aquifers associated with energy resources could be added to state or federal statutes. General water quality information is now required in the Environmental Impact Statement process for most major energy development projects. However, specific predevelopment baseline data is not required in the permitting process for mining, drilling, or energy conversion facilities. Such predevelopment monitoring could be added as a permitting, leasing, or statutory requirement.

The costs of such a baseline data collection effort will depend on several site-specific factors. For example, with groundwater it will depend on the number, character, depth, and extent of aquifers. Monitoring costs would represent only a small portion of the total costs of a mine or energy conversion facility. Both because groundwater quality does not fluctuate rapidly and because a lengthy predevelopment period of measurements is not feasible, frequent groundwater measurements would not be taken over an extended period of time. In contrast, collection of data on surface waters should extend over the longest time period feasible.

Water Quality Control Plan

A water quality control plan covering both surface and groundwater would be part of a regulatory program that specified the existing hydrologic resources, potential disturbances, control measures, monitoring practices, and restoration options for all energy development projects. This plan would be submitted by energy developers to the relevant local, state, or regional resource management agency.

A water quality control plan would be designed to protect hydrologic resources within the geographic boundaries of a developer's resource holding, but would also apply outside the "fence line" or area of resource ownership. This plan would provide information to planners and government officials for assessing potential effects of aggregated development on the regional hydrologic system.

As discussed in Chapter 3, such a plan is already required for coal surface mining operations under the Surface Mining Control and Reclamation Act (SMCRA) of 1977 and under various state laws. To obtain a permit, the SMCRA basically places the burden of proof on the mining company to show that groundwater quality will not be significantly harmed. However, no similarly comprehensive groundwater protection plans are presently required for extraction of the other five resources we have considered or for disposal of wastes from energy processing and conversion facilities. As noted in a recent Colorado study,

> one of the greatest needs in the area of water quality management for active mining operations is for regulatory coordination and program consistency between the MLRB [Mined Land Reclamation Board], the WQCD [Water Quality Control Division, Colorado Department of Health], and the federal land management agencies [Colo., Dept. of Health, WQCD 1979, p. 24].

A plan, such as the one described here, should be comprehensive enough to minimize water quality degradation for all energy resources and for both extraction and conversion activities.

For mining, data and performance requirements in this water quality control plan could be modeled after the permit requirements for surface coal mining (as described in Chapter 3). The plan could also contain sections similar to the effluent discharge plan now required by the state of New Mexico. The New Mexico plan is intended to provide information on the fate of pollutants within or adjacent to energy developments which may not come under the National Pollutant Discharge Elimination permit system; that is, discharges that may not enter navigable or interstate waters. In general, the water quality control plan would include identification of potential water pollution from both point and nonpoint sources and would project the effects on surface and groundwater quality.

For both mining and waste disposal, the technological measures available are in the early stages of development and testing. Because these technologies generally are both labor- and materials-intensive with significant capital and operating costs, implementation strategies need to be carefully designed. For example, in one demonstration project on the handling of overburden, it was estimated that the selective burial operations cost from 1.1 to 1.5 times as much as the normal operations of a surface mine (Dollhopf et al. 1977).

Implementation of control measures by specifying design standards may have considerable disadvantages such as restricting options or requiring technical specifications that are inappropriate for some locations. Implementation based on technical review and performance standards would minimize delay, excessive costs, and possible deployment of inappropriate controls. Efficient implementation could be achieved by requiring review of groundwater protection techniques and restoration options and the development of appropriate site-specific plans based on performance standards. Examples of these general performance standards would include:

(1) Use of best practicable surface and subsurface mining practices to minimize change in groundwater quality;

(2) Specification of the maximum permissible seepage from waste disposal areas (allowing the industry to choose, for example, the amount of pretreatment or the type of pond liner required to meet the standard);

(3) Removal of contaminated groundwater where feasible to enhance the immediate beneficial use of aquifers; and

(4) Use of best practicable aquifer restoration technology to preserve the long-term beneficial use of groundwater.

In addition to increasing energy production costs, comprehensive controls could slow the pace of development because of the planning requirements and because some controls may require procedures which in themselves slow operations. For example, the selective handling of overburden layers, depending on their toxicity, usually requires truck and shovel operations instead of draglines. Because of these factors, industry resistance can probably be expected. However, the comprehensive planning described here represents an extension of existing controls for coal surface mining, and its application to all energy development activities would ensure a higher degree of protection for the West's water resources.

Monitoring and Research

An expanded monitoring and research program would lead to a better understanding of the risks to water quality, especially groundwater quality, from energy development in the West. At the present time, publicly available data on the quality of groundwater in many areas are limited. The U.S. Geological Survey (USGS) does have an active and broad-based monitoring program for both surface water and groundwater, so new or expanded monitoring programs should be coordinated with and built upon the USGS system. Monitoring is being conducted by the energy industry, but these data are largely proprietary. The state of New Mexico has established periodic monitoring requirements for all surface and subsurface discharges, but other states in our eight-state study area do not have this comprehensive monitoring requirement.

The cost of monitoring is largely dependent on the number of wells, the frequency of sampling, and local conditions. A major sampling program could range in cost from $100,000 to $200,000 per year for a single mine or waste disposal site, depending on well investment, personnel, and analytical equipment needs.

In terms of research, there are already energy-related groundwater quality research programs within various federal agencies (primarily the Department of Energy, EPA,[1] and USGS), state governments, and universities. However, the current level of research is not adequate to provide sufficient and timely information. To be most effective a wide range of studies is needed to examine problems connected with the several energy resources in a variety of environments, with primary emphasis on coal, oil shale, and uranium. These studies should examine potential risks and alternative controls for mining, in situ recovery, and waste disposal.

Examples of the types of studies needed for holding ponds are listed below:

- Failure rates of holding ponds: Historical data need to be gathered on the kinds of failures, their significance, and the critical variables, including rainfall, floods, and structural failure. These data are not now accessible.

- Movement of pollutants: Field and laboratory research is needed on the migration of pollutants through various types of pond liners during normal operation and over the long term. Studies by the EPA and the USGS have begun to assess this problem. Coordinated studies in a variety of western locations need to be initiated. In addition, studies are needed on the transport and fate of pollutants in soil and groundwater systems after leaching from holding ponds.

- Water consumption by holding ponds: Holding ponds may increase water consumption of energy facilities by 10 to 20 percent because of evaporation of water that, through the use of costly technologies, could have been cleaned, treated, and reused. Information on the process design requirements and economic costs of these alternative technologies could be made more available to decisionmakers.

Generally, these holding pond studies reflect the need to apply technological criteria to the unique water resources and environmental conditions found in the Rocky Mountain and Northern Great Plains states. Current information problems include: inadequate historical documentation of problems; inadequate attention to technical alternatives for achieving standards for discharge of pollutants and their associated environmental and economic trade-offs; and inadequate dissemination of existing information to local, state, and federal decisionmakers.

Of course, the major obstacle to such a comprehensive research program is the cost. While the costs are difficult to estimate and would depend largely on the scale of effort, a moderately sized program of the type being described here would cost between $20 million and $50 million over the next ten years. Much of this research would probably require federal government support, but to be most efficient it should be undertaken cooperatively with states and energy developers.

TEMPORARY WASTEWATER TREATMENT MEASURES

As discussed in Chapter 3, western communities near large energy development projects will experience rapid population increases. After construction is completed, the population may decrease because less manpower is needed for operation, or the population may continue to increase as additional energy projects and associated activities are located in the area. Because wastewater treatment plants in many western communities are already at or near capacity, a rapid and relatively large population increase is likely to result in inadequately treated sewage. Also few small communities in the West affected by energy development will be able to afford their share of the cost of upgrading and/or expanding sewge treatment facilties to meet the requirements of the Clean Water Act (1977). The need is for a less costly, yet effective, method to rapidly expand wastewater treatment capacity. The policy alternative considered here would allow the use of sewage lagoons (or waste stabilization ponds) as a temporary measure until more conventional secondary treatment facilities can be constructed.

Waste stabilization ponds are currently used by many communities in the West, especially the smaller ones. Their main advantages include low capital and maintenance costs and simplicity of operations. Although both labor and capital costs for constructing and operating waste stabilization ponds are very small compared to conventional secondary treatment facilities, both require wastewater conveyance systems which may cost anywhere from $176 to $1,195 per capita (1976 dollars) in small communities (Dames and Moore 1978). The capital costs for construction of waste stabilization ponds include land, wastewater conveyance systems, and construction of berms. Generally, local manpower and skill is sufficient both for construction and for operation. Land is readily available in most areas of the West where the energy-related boomtown conditions are likely to occur. Often, however, the most desirable location for a pond is also desirable farmland. While conventional treatment facilities would also compete for farmland, they typically require less land than ponds.

In addition to lower costs, ponds can be designed and constructed much more rapidly than conventional secondary treatment facilities. Also since capital costs for ponds are less, if energy development does not occur as expected the community will not have lost as big an investment. And, waste stabilization ponds are more readily decommissioned than conventional systems; the land for sewage lagoons can be easily drained, filled in, and converted back to its original use in most cases.

The major disadvantage of waste stabilization ponds compared to conventional wastewater treatment facilities is their apparent inability to meet EPA's secondary sewage treatment standards throughout the year. In the West difficulties may occur with excess suspended solids during the late summer and anaerobic conditions during deep ice cover in the winter. In fact, in 1979 EPA and several western states relaxed the suspended solids limitation for sewage lagoons to allow their continued use. However, in many areas of the West where net evaporation rates are high and the necessary land is readily available, waste stabilization ponds can often be constructed to retain all of a municipality's wastewater without having any discharge (as is encouraged in Utah). In northern and higher elevation areas of the West, ice cover limits the effectiveness of lagoons during the winter. In such areas, ponds may have to be quite large to retain the wastewater during these periods (as is currently required in North Dakota). Of course, a major cost of complete retention of wastewaters is the increased consumptive use of water. For example, as discussed in Chapter 3, a 400,000 barrel per day (bbl/ day) oil shale industry was estimated to result in an increased municipal wastewater flow during construction of three million gallons per day (MMgpd) (or 3,000 acre-feet per year [AFY]). Thus, the use of sewage lagoons with complete retention would increase the water consumption associated with energy development by this amount.

Another consideration in the use of sewage lagoons is that if the ponds are not constructed while temporary and rapid growth occurs, existing facilities are likely to be overloaded or by-passed altogether, resulting in even greater pollution of receiving waters. Requiring capital-intensive facilities before or at the time of the population increase puts a large financial strain on a community already struggling to provide sufficient services. The construction of ponds, at least on a temporary basis, would allow small western communities to provide economical wastewater treatment while collecting the necessary front-end financing from the expanded tax base for more conventional treatment facilities.

SALINITY CONTROL PROJECTS

Description

Much of the salt from man-made or natural sources may not originate at point sources or may be difficult to manage by point source controls. Therefore, salinity control alternatives can include desalination plants, diverting flows around areas of high salt pick-up, eliminating saline flows by evaporation (including agricultural return flows), and improving irrigation practices. These approaches are currently being emphasized to control salinity in the Colorado River. The Colorado River Basin (CRB) Salinity Control Act (1974) established a comprehensive program, including authorization for the construction of five salinity-control projects and the study and planning of twelve other projects. The projects authorized for construction by the CRB Salinity Control Act are described in Table 6-1. Table 6-2 shows the projects studied or planned by the Water and Power Resources Service (WPRS) (now redesignated the Bureau of Reclamation) under the CRB Salinity Control Act; these projects have not yet been approved for construction.

Desalination techniques include chemical pretreatment and several methods for separating dissolved solids from water, such as freezing, distillation, and filtering through membranes to remove brine (reverse osmosis). For example, the salinity control complex being constructed near Yuma, Arizona, which was authorized by the CRB Salinity Control Act (1974), includes a desalination plant that will use a combination of chemical pretreatment and membrane systems. Now scheduled for completion in the mid-1980's, this plant would treat about 107,500 AF of saline agricultural runoff each year and remove about 515,000 tons of dissolved solids per year (U.S., DOI, WPRS 1980). Brine discharge from this plant would be diverted from Yuma, Arizona, to a drainage channel terminating in the Gulf of California. In other locations distant from marine water, brines could be disposed of by evaporation or by pumping to selected groundwater formations. The Yuma desalting project will have no effect on salinity above Imperial Dam since it is located near the Mexican border. It is included here for comparative purposes since similar desalting plants have been discussed for the Colorado River above Imperial Dam.

The practices of diverting streams around areas of high salt pick-up or intercepting highly saline groundwaters can also control salt loading. In fact, several diversions are included in plans for the Colorado River as shown in Tables 6-1 and 6-2. Impoundments are used to prevent highly saline waters from flowing into a stream in which salinity is being controlled, and canals are used to circumvent saline seeps or formations. This option is being considered for the LaVerkin Springs Unit, the Price River Unit and several others of the Colorado River Water Quality Improvement Program (CRB Salinity Control Act, 1974, Title II).

The technique of eliminating saline flows by constructing storage ponds and allowing the saline water to

TABLE 6-1:   Salinity Control Projects Authorized for
Construction Under P.L. 93-320

Project	Description	Current Status
Yuma Unit, Arizona	The project includes a desalting plant, construction of a bypass drain and lining a canal. Other activities include increased irrigation efficiency, acreage reduction, and protective groundwater wells along Mexican border.	Size of the desalting plant has been reduced to 96 MMgpd; construction of various component has been delayed.
Paradox Valley Unit, Colorado	Groundwater pumping will lower the freshwater brine interface below the river channel; pumping to an evaporation pond is being considered to remove about 180,000 tons of the 205,000 tons of salt that annually seep into the Dolores River.	Test wells have been started; data for design are being collected.
Grand Valley Unit, Colorado	Improved irrigation distribution systems (canal and lateral lining) and management to increase efficiency are planned to reduce the salt entering the Colorado River by 410,000 tpy.	Preliminary investigations are completed; inital phase monitoring is to be established.
Crystal Geyser Unit, Utah	This abandoned oil test-well contributes about 3,000 tpy salt; the flow was to be collected and evaporated.	Construction has been deferred indefinitely due to low cost-effectiveness.
Las Vegas Wash Unit, Nevada	Originally, interception of groundwater and evaporation was planned for Stage I and a desalting plant for Stage II. Lining of ponds in the area has resulted in a changed hydrology.	Construction has been delayed pending further study.

Source: U.S., DOI, BuRec 1979, pp. 71-85; U.S., GAO, Comptroller
General 1979, pp. 35-45.

TABLE 6-2: Projects for Salinity Control Being Studied
Under Title II of P.L. 93-320

Project	Description
Irrigation-Source Control Projects	
Lower Gunnison Basin Unit, Colorado	Irrigation management and improved water systems are planned for 160,000 acres contributing about 1.1 million tons of salt per year. Initial program has begun.
Uinta Basin Unit, Utah	Irrigation management and improved water systems are planned for 170,000 acres contributing about 450,000 tons of salt per year. Initial program has begun.
Colorado River Indian Reservation Unit, Arizona	Irrigation management and improved water systems are planned for about 80,000 acres. A study found no salt is contributed to the Colorado River by the reservation and development of the project has been discontinued.
Palo Verde Irrigation District Unit, California	Irrigation management and improved water systems are planned for about 91,400 acres contributing about 152,000 tons of salt per year.
Point-Source Control Projects	
LaVerkin Springs Unit, Utah	River flow will be diverted around the springs; the saline water will be pumped to a desalting plant. About 103,000 tons of salt per year would be removed from the stream.
Lower Virgin River Unit, Nevada (replaces Littlefield Springs Unit)	Wellfield or barrier dam and solar evaporation to collect irrigation return flow and saline groundwater are planned.
Blue Springs Unit, Arizona	Feasibility studies to control 160,000 AFY that contribute about 550,000 tpy of salt were planned but, due to high costs and environmental problems, these plans have been cancelled.

(continued)

Table 6-2: (continued)

Project	Description
Glenwood-Dotsero Springs Unit, Colorado	Treatment to remove most of the 250,000 tpy of salts contributed to the river is planned.

Diffuse-Source Control Projects

Project	Description
Big Sandy River Unit, Wyoming	Interception of saline groundwater seeps is planned to remove about 80,000 tons of salt per year in 6,000 acre-feet. Evaporation in a pond is being considered.
Price River Unit, Utah	Selective withdrawal, farm management, and other improvements in the water systems are planned to control 100,000 tpy of salt.
San Rafael River Unit, Utah	Selective withdrawal, farm management, and other improvements in the water systems are planned to control 80,000 tpy of salt.
Dirty Devil River Unit, Utah	Selective withdrawal, farm management, and other improvements in the water systems are planned to control 80,000 tpy of salt.
McElmo Creek Unit, Colorado	Planning has not progressed far enough to identify potential alternatives.
Meeker Dome Unit, Colorado	Collection of salt wells and seeps with treatment or utilization is being considered.

Source: CRB Salinity Control Forum 1978, pp. 42-49; U.S., DOI, BuRec 1979, pp. 85-99.

streams and to saline agricultural return flows; however, many large ponds would be required. This is nevertheless a viable option in some cases and this approach is planned for saline streams in the San Rafael and Dirty Devil River units of the CRB Water Quality Improvement Control Program.

As shown in Tables 6-1 and 6-2, there are several projects to reduce salinity contributions from irrigated agriculture. These include the Grand Valley, Lower Gunnison Basin, Uintah, Colorado River Indian Reservation, and Palo Verde Irrigation District units. Activities planned to reduce salinity contributions from these sources include improved irrigation management and improvements in the water systems. Irrigation management activities are nonstructural and include such things as scheduling the delivery and application of water. Structural actions to improve water systems include canal lining, use of pipe systems, and land forming. The effect of such actions on water availability is discussed in Chapter 4.

How Much Can Salinity Be Reduced?

It is not meaningful to project the total reduction in salinity likely to occur if all of the projects authorized for construction were in operation because several of them have been delayed indefinitely or are being substantially revised. However, as shown in Table 6-3, the effects of some presently planned salinity-control projects can be projected. The combined effect of these seven projects alone would be an estimated 70.2 milligrams per liter (mg/l) salinity concentration reduction at Imperial Dam (U.S., DOI, BuRec 1979). Another study has shown that the Grand Valley Unit will reduce salinity at Imperial Dam by about 43.0 mg/l (U.S., GAO, Comptroller General 1979, pp. 29-30). These projects have been chosen largely because they are thought to be the most cost efficient and effective in controlling the major sources of salinity in the basin. Additional projects may be more expensive in proportion to their effect on salinity at Imperial Dam.

What Are the Costs?

Along the southern part of the Colorado River, constructing desalination plants and evaporating highly saline waters can remove a ton of salt for approximately $30 to $40 (U.S., DOI, BuRec 1978, pp. 195-211). The costs of flow containment or diversion can vary greatly depending on the specific project. For example, the Paradox Valley Program, which includes evaporation of

TABLE 6-3: Estimated Salt Load and Salinity
            Concentration Reduction by Selected
            Salinity Control Projects

Project	Estimated Salt Removal (tons)	Estimated Salinity Salinity Reduction at Imperial Dam (mg/l)
Paradox Valley Unit, Colorado	180,000	18.2
Uintah Basin Unit, Utah	100,000	10.0
LaVerkin Springs Unit, Utah	103,000	9.0
Big Sandy River Unit, Wyoming	80,000	7.0
Price River Unit, Utah	100,000	10.0
San Rafael River Unit, Utah	80,000	8.0
Dirty Devil River Unit, Utay	80,000	8.0

Source: U.S., DOI, BuRec 1979, pp. 85-99.

highly saline water, is estimated to cost about $10 per
ton (1975 dollars) of salt removed (Utah State U., Utah
Water Research Lab. 1975, pt. 2, p. 262).

Table 6-4 shows estimates of the costs per mg/l salinity reduction at Imperial Dam for four of the major
projects originally authorized by the CRB Salinity Control Act (1974). Although considerable uncertainty
exists about reductions and their associated costs, this
measure is more relevant for comparison than simply estimating the total tons of salt removed or the cost per
ton. As indicated, these annual costs range from about
$200,000 to $1 million per mg/l salinity reduction.
These costs are for some of the area's largest sources
of salinity; thus they may not be indicative of the
costs to be expected for controlling other sources of
salinity.

An additional cost of some of these projects will be
loss of water in the Colorado River. For example, wells
for the Paradox Valley unit will remove about 3,620 AFY
from the river which will be pumped to an evaporation
pond. The LaVerkin Springs unit in Utah is planned to
include pumping about 2,300 AFY of brine to an evaporation pond. The Big Sandy River unit might evaporate
about 6,000 AFY as brine in a pond (U.S., DOI, BuRec
1979, pp. 71-99).

TABLE 6-4:   Estimated Costs of the Four Projects
             Authorized by Title II of P.L. 93-320[a]

Project	Salinity Reduction at Imperial Dam (mg/l)	Annual Equivalent Cost	Annual Equivalent Cost per mg/l Salinity Reduction at Imperial Dam
Paradox Valley	18.2	$ 3,507,000	$192,700
Grand Valley	43.0	10,824,000	251,700
Las Vegas Wash[b]	9.0	8,727,000	969,700
Crystal Geyser[b]	0.3	234,000	780,000

[a]U.S., GAO, Comptroller General 1979, pp. 29-30. The date of these cost estimates was not provided.

[b]These units have been delayed indefinitely but are included for comparative purposes to estimate the range of costs likely to be found in the salinity control projects.

Because these technological controls are all capital-intensive, they will almost certainly require public (largely federal) financing. Thus, taxpayers living in regions other than the West will pay a large proportion of the costs. Because agriculture is the major man-made cause of salinity and since lost agricultural production is a major cost of high salinity levels, then a significant share of such expenditures can be reasonably viewed as a subsidy for irrigated agriculture. While schemes might be developed to assess water users for the cost of the project, such strategies would be difficult to implement and, unless carefully designed, could cause downstream users to pay for problems largely originating upstream.

Westerners will be the primary benefiaries of improved surface water quality. For example, agricultural benefits of each mg/l decrease have been estimated to range from about $46,000 to $108,000. Municipalities are estimated to benefit by about $120,000 for each mg/l reduction in salinity by, for example, not having to replace appliances and water treatment and distribution facilities as frequently. Benefits to industry are estimated to be about $1,500 for each mg/l decrease in salinity (Utah State U., Utah Water Research Lab. 1975, pt. 1, pp. 166-68). Thus, the total economic benefit was estimated to be between $167,500 and $229,500 for each mg/l salinity improvement (1975 dollars). In early 1980 WPRS updated this figure to $343,000 per mg/l

improvement (U.S., DOI, WPRS 1980). Environmental benefits include the protection of fish and other aquatic
life as well as enhanced recreational uses of these
waters.

Are There Implementation Obstacles?

Salinity control projects are generally constrained
by the high capital costs, particularly for desalination
plants. Nevertheless, the legal and regulatory basis
has already been established for each of these alternatives, so implementation may be less difficult than
for many other ways to control the sources of salinity.
For example, as discussed in Chapter 3, four salinity
control projects were originally authorized by the CRB
Salinity Control Act (1974), and commitments were made
to twelve others if planning reports were favorable.
Hence, this alternative could be implemented in the CRB
with existing programs and agreements. However, these
projects have been the subject of considerable controversy, and they have all encountered delays. The major
opposition has arisen from parties-at-interest who feel
that the environment will be harmed by some of the specific projects and from others who feel that the benefits do not justify the large public expenditures or
that more cost-effective methods are available (U.S.,
GAO, Comptroller General, 1979).

SALINITY OFFSET POLICY

Description

The general purpose of a salinity offset policy
would be (1) to prevent any further increases in salinity concentrations or loadings in streams with high
salinity levels, or (2) to reduce salinity concentrations or loadings in streams that are above desired levels. Although a salinity offset policy could be implemented comprehensively, for the purposes of this study
it will be discussed only as it applies to new energy
development activities. Under this policy, a proposed
development which adversely affects salinity would be
allowed to proceed only if some other offsetting activity were implemented, so that net salinity levels would
not increase. For example, if a proposed energy facility in the CRB would result in a salinity concentration
increase as measured at Imperial Dam, then salinity from
other sources would have to be reduced by at least an
equivalent amount. This could be accomplished in any
way the developer desires; presumably, the least costly

salt reduction method available would be selected.
Projects to offset salinity could include: improving
agricultural irrigation methods to reduce salinity in
return flow for a particular irrigation district; reducing a natural source of salinity; improving a municipal wastewater treatment plant to reduce salinity discharges, and funding part of a desalination plant. In
addition, energy developers could purchase irrigated
land and the appurtenant water rights. The water would
be permitted to flow unused downstream, thereby eliminating the saline irrigation return flows from the land
and compensating for the salinity effects of the energy
facility.

Such a policy need not necessarily be limited to
energy development. Since energy development is a much
smaller contributor to total salinity than natural and
agricultural sources, a less restrictive program could
be more effective and it would be more equitable than
having it apply only to energy industries. Nevertheless, implementing the policy for just the energy industry would not be unreasonable, in that it would be designed simply to offset the salinity increases caused by
energy. Salinity problems due to other sources could
still be dealt with using whatever policy approach seems
most reasonable for that source, including an "offset
policy."

The Need for Improved Salinity Models

Salinity/flow relationships are complex and not well
understood. For example, the effects at Imperial Dam of
a discharge of 100 tons of salt in the San Juan River in
New Mexico would probably not be the same as an equal
discharge in the Yampa River in Colorado. The effects
of water withdrawals in different portions of the basin
would also not be equal. Therefore, one of the difficulties associated with this alternative is the inac-
curacy and uncertainty of predictive salinity models.
Current models may overestimate or underestimate salinity effects by as much as 100 percent. This could result in more strict controls and higher costs than necessary. On the other hand, salinity levels could
increase if the effects of an energy facility are under-
estimated or if the effectiveness of an offset policy is
overestimated.

For this alternative to be effective, a standard
salinity model capable of predicting the effects of
individual actions, independent of other water use decisions, is needed. Even if the model is not completely
accurate, adoption of the policy by all states in a
basin and use of a common model would minimize

inequities among developers. Each project would have to compensate for its own estimated salinity effect.

For salinity effects of discharge and of reduced flows, some believe that models could soon be formulated for the CRB that would show the effects of a single action. Others feel current models are too inaccurate and make over-simplifying assumptions. New models based on salinity transformations that occur in the river system are needed. In the Upper Missouri River Basin salinity models are not currently well enough developed to predict the effect of a particular withdrawal or discharge. The refinement of existing salinity models and the development of models for other pollutants is needed before this alternative could be effectively and equitably implemented.

How Could the Policy Be Implemented?

Establishing a salinity offset policy for energy development would face serious legal, institutional, and political barriers. However, by limiting the policy to the energy industry far fewer constraints would exist than if the policy were applied to all new water development activities.

To be most effective, a salinity offset policy would need to be adopted by each of the states in a river basin. Adoption by one or only a few of the states might create conflicts with programs of the other states, resulting in fragmented management and fewer offsets than would be available if the policy were adopted by an entire basin. States themselves could not fully evaluate the offset plans proposed by developers within its boundaries but would have to use the agreed upon basin-wide predictive salinity model to estimate the net effects at some specific location, such as Imperial Dam or the main stem of the Yellowstone River. Interstate cooperation and coordination would be needed for those projects where the offset occurs in another state.

The legal and institutional basis for establishing a salinity offset policy may already exist in the CRB Salinity Control Act and the CRB Salinity Control Forum. However, such a policy requires linking water allocation decisions with water quality standards, which is not part of the current water allocation system in western states. The types of changes that would be required in the water laws of each state are unknown at this time and more research into this question is required. One attempt has been made at establishing a salinity offset policy in Colorado. The Northwest Colorado Council of Governments (NWCCOG) Draft Areawide Water Quality Management Plan (208 plan) encompassing six counties on the West Slope included a requirement that

```
SALINITY STANDARDS CONFLICT

        During public hearings by the Colorado Water
Quality Control Commission over salinity standards
and their implementation an EPA representative stated
the agency believes Colorado may be contemplating
"a serious departure from the basin-wide approach
...[to the] joint salinity control effort" by the
Colorado River Basin Salinity Control Forum.  Others
including spokesmen for the Northwest Regional
Council of Governments, the Northern Colorado Water
Conservancy District and Chevron Oil Shale Company
voiced divergent views.  Some contended it was EPA
who was trying to go back on its own acceptance of
the salinity control implementation plan of the
Forum.                                --Tweedell 1980
```

...future new municipal and industrial water projects from the natural drainage of the Colorado River within [the] Planning and Management Region...offset salinity increases by mitigation measures designed to ensure that the cost of the present and prospective uses of water and land within the natural drainage of the Colorado River within the State of Colorado shall not be increased (Comarc Design Systems, Wallace McHarg Roberts & Todd and NWCCOG, 1978).

Objections to this plan were raised mainly by present and potential future transmountain diverters, such as the Denver Water Board (DWB) and the Northern Colorado Water Conservancy District and municipal subdistricts (Lane 1979; Gill 1979a, 1979b). They contended that the offset policy, which would limit withdrawals and diversions from the area, is beyond the legal authority given the NWCCOG by state and federal legislation. The DWB and others filed suit in state district court to block the plan. Eventually, Governor Lamm vetoed the salinity offset portion of the plan, fearing it would inhibit the development of Colorado's remaining share of Colorado River water. The state must now devise an alternate method to control salinity increases.

One of the practical difficulties in carrying out a
salinity offset policy is the constraint on developing
offsets in agriculture, which is a major source of salinity. As described in Chapter 4, substantial reductions
in salinity loads are possible by changing water use
practices in agriculture (e.g., by using sprinklers or
by lining canals). In addition to being relatively expensive such water saving changes are discouraged by
current state water laws, principally the nonimpairment
and beneficial use doctrines. Under the nonimpairment
doctrine a water user cannot impair the rights of others through a change in water use. An energy facility
seeking a salinity offset by reducing saline return
flows from irrigated agriculture by paying for a sprinkler irrigation system, for example, could not legally
reduce the return flows if the rights of users who depend on the return flows would be harmed. State laws
also require that water be put to a beneficial use. If
the salinity offset plan of a developer results in less
water use for a specific purpose, the water user may
lose the portion of the water right that is no longer
used.

What Are the Economic Costs?

The cost to an energy producer of a salinity offset
would depend on the level of salt loadings, the amount
of water withdrawn from the river system, the location
of the withdrawal, and the salinity offset planned.
Three proposed energy developments will be used to estimate the cost of salinity offsets to potential developers: an oil shale facility in Colorado; two coal-
fired power plants in Northwest Colorado; and a coal
gasification facility in New Mexico. A 47,000 bbl/day
oil shale conversion plant has been proposed in Colorado. This facility would consume about 7,200 AFY from
the Colorado River near Grand Valley, which would result
in a projected average increase of 0.12 mg/l in salinity
below Hoover Dam (U.S., DOI, BLM 1976a, vol. 1, pp.
IV-97 through IV-101). In the previous section on constructing salinity control projects, cost estimates for
the four salinity control projects authorized by the CRB
Salinity Control Act ranged from $200,000 to $970,000
per mg/l decrease in salinity concentration at Imperial
Dam. Using this range as a guide, an increase of 0.12
mg/l would result in an annual cost for salinity offsets
for the oil shale facility of $24,000 to $116,000. Two
additional coal-fired power plant units totaling 760
megawatts have been considered for Craig, Colorado (Units
3 and 4). These units would consume about 12,000 AFY
and result in a projected average salinity increase of

0.7 mg/l below Hoover Dam (U.S., DOI,BLM 1976b, pp.
III-3 through III-9). Assuming the same salinity control costs as above, the annual costs for salinity offsets would be from $140,000 to $680,000. Finally, a 410
million cubic feet per day coal gasification plant proposed for San Juan County, New Mexico, would consume
15,000 AFY and result in a projected annual average salinity increase at Imperial Dam of 2.4 mg/l (U.S., DOI,
BuRec 1977, vol. 1, pp. 3-22 through 3-27). Offsets for
this increase might cost from $480,000 to $2,330,000
each year.

Costs for these salinity offsets, summarized in
Table 6-5, will be added to the costs of energy production, some or all of which will eventually be passed on
to consumers. If synthetic gas costs $4.50 per thousand
cubic feet (Mcf), then the cost for salinity offsets for
the gasification plant would result in a cost increase
of at most 0.44 percent. For electric power generation
salinity offsets would increase costs less than about
0.01 cents per kilowatt hour (kWh). This would be an
increase of about 0.25 percent if electrical generation
costs are 4 cents per kWh. For the oil shale case with
produced oil valued at $35 per barrel, salinity offsets
would increase costs less than 0.02 percent.

As these numbers indicate, these costs are relatively small. Generally a salinity offset policy would
promote economic efficiency since energy companies could
utilize the lowest cost offsets available. In addition,
actions may be taken to decrease the salinity impacts of
a proposed facility and thereby reduce the amount and
cost of offsets. Salinity impacts can be reduced when
poorer quality water is used. For example, if six 250
MMcfd coal gasification plants located in the Four Corners area of New Mexico withdraw water between Farmington and Shiprock instead of upstream between Bloomfield
and Farmington, the salinity increase would be about one
mg/l less measured at Imperial Dam. This benefit would,
of course, have to be balanced against possible increased treatment costs for the lower quality water and
increased transportation costs. In some situations, if
an energy facility were to use highly saline surface
waters, salinity benefits could result (see Chapter 5).
In this case, the company could receive a "salinity
credit" which could then be sold to other companies
needing a salinity offset. A second way energy facility
salinity impacts may be mitigated is to reduce water requirements by improving process design or using less
water consumptive cooling technologies, as discussed in
Chapter 4. For example, a combination of wet and dry
cooling for a gasification plant in New Mexico could
reduce water withdrawals by about 16 percent compared to
all wet cooling.

TABLE 6-5: Summary of Salinity Offset Costs Estimated
for Three Example Energy Facilities

Facility	Projected Salinity Increase (mg/l)	Annual Cost of Salinity Offset[a]	Offset Cost ($/product)	Cost as a Percentage of Product Price[b]
Oil Shale Conversion (47,000 bbl/day)	0.12	$24,000-116,000	<0.01/bbl	<0.02%
Electric Power Generation (760 MWe at 80% load factor)	0.70	$140,000-680,000	<0.0001/kWh	<0.25%
Gasification (410 MMscfd)	2.40	$480,000-2,330,000	<0.02/Mcf	<0.44%

MWe = megawatt-electric   MMscfd = million standard cubic feet per day

[a]Based on an estimated annual cost range of $200,000 to $970,000 per mg/l of salinity.

[b]Based on the following energy product costs: shale oil $35/bbl; electric power $0.04/kWh; and synthetic gas $4.50/Mcf.

Additional public sector costs of the salinity offset policy will result from the need for increased manpower to review permits and enforce permit requirements
in each state. Also, there will be costs in developing
and refining basin-wide salinity models to predict the
effects of consumptive uses.

What Are the Major Barriers?

Energy developers are not likely to favor this alternative because of the increased costs and the possible
procedural delays in finding offsets. However, if salinity standards are a barrier to gaining water rights or
project permits, any alternative which allows development or makes clear what is needed for project construction and operation may be attractive. Environmental
interest groups are likely to favor any option to reduce
salinity but may question the adequacy of the salinity
models and the increased public cost of administration.
If a salinity offset policy is not to be a procedural barrier to energy development, close cooperation
will be needed among energy developers, potential

sources of salinity offsets, and administrative agencies. At the present time there is no administrative mechanism to identify potential salinity offsets (either natural or man-made) and to promote cooperation between potential sources of man-made offsets and energy developers. For example, many agricultural interests, including irrigation districts, may be reluctant to assist energy developers in obtaining offsets without some incentive from the state or federal government. Without this cooperation developers will have difficulty identifying and obtaining adequate offsets.

One way that basin states could assist with this alternative would be by a regional banking of salinity offsets. In other words, the CRB Salinity Control Forum could identify and clean up salinity sources within its jurisdiction and use this to attract new development. Part of the condition for locating in that basin could be that the industry pays for the salinity offset that has already been completed. This would shorten siting time and regulatory complexities for industries.

SUMMARY AND CONCLUSIONS

Energy development can contribute to water quality problems in three primary ways:

- *Contamination of surface and groundwater due to land disturbance and waste disposal;*

- *Increased salinity concentration downstream due to consumptive use of water; and*

- *Inadequate sewage treatment in energy boomtowns.*

This chapter has described and evaluated four policy options for dealing with these water quality problems. Table 6-6 lists the alternatives and summarizes some of the key findings for each. Because water quality and water availability problems are often closely linked, several of the policy alternatives considered in other chapters can also have important implications for water quality. Especially important in this respect are the options for using saline water in energy facilities (Chapter 5) and for improving the efficiency of water use in agriculture (Chapter 4).

A mix of the four policy options considered in this chapter could be used to contribute to the protection of western water quality. The first alternative, improved water quality controls for energy development activities, consists of three primary elements: predevelopment

TABLE 6-6: Summary and Comparison of Water Quality Protection Alternatives

Alternative	Current Status	Advantages	Disadvantages	Feasibility
Improved water quality control	Adequacy varied among resources and technologies; groundwater quality control plans for coal mining required; generally other elements inadequate	Needed baseline data provided by predevelopment monitoring; increased project costs relatively small Minimization of long-term, almost irreversible effects on groundwater quality by overall program	Somewhat increased costs and slower development ment of energy Monitoring and research costs large; federal funds probably required for majority of effort	Basically presents expansion and coordination of existing programs Possible public and industry opposition due to increased energy costs and federal expenditures
Temporary sewage treatment	Sewage lagoons already being used in some western communities	Low capital and operating costs Short lead-time May reduce pollution if inadequate treatment or by-passing would otherwise occur Easily reversible	May not meet discharge standards Increased water consumption due to complete retention	Would require revisions of official EPA policy Likely increased local support due to lower initial costs
Salinity control projects	Current approach to salinity control in the CRB	Can remove large amounts of salts; e.g., salinity reduced by 9, 18, and 43 mg/l at Imperial Dam by three authorized projects Economic benefits estimated at approximately $343,000 per each mg/l reduction Environmental benefits from salinity reduction	Large capital and operating costs; annual costs between $0.2 and 1.0 million per mg/l reduction in salinity Will reduce river flow; e.g., 3,620 AFY consumed by Paradox Valley Unit; 8,200 AFY consumed by Las Vegas Wash Unit Environmental risks from some proposed projects	Institutional and regulatory basis already in existence Some opposition due to large federal cost outlays required and the environmental impacts of some projects

Table 6-6: (Continued)

Salinity offset policy for energy development	Not currently used	Could control salinity concentration increases for energy due to water consumption	Salinity models not well developed	Opposition reduced by limiting policy to energy development
		Economically efficient since lowest cost offsets could be used	Could slow energy development; however, acceleration possible by states "banking" offsets	Likely opposition due to increased energy costs and possible regulatory complexity
		Increased energy costs probably small; e.g., between 0.02 and 0.5 percent		Legal basis uncertain

monitoring of water quality, water quality control plans, and monitoring and research. These options are an expanded and more comprehensive approach to already ongoing activities, not a radically new program. The goals of this option are to ensure that best available practices are used in the short-term and that over the longer term the knowledge about geohydrology and the behavior of potential contaminants is improved. This will ultimately improve our ability to define the risks of energy development and the effectiveness of various control measures. Without such a program, chances will increase that serious, and in some instances irreversible, damage will occur to the West's water resources, especially groundwater. The disadvantages of this option are that energy costs could be increased due to stricter control measures and that increased public sector expenditures will be required for research. However, given the high level of uncertainty concerning potential impacts and the likelihood of expanded production of western energy resources, the costs appear to be worth paying.

The alternative considered for dealing with municipal wastewater problems would allow the use of sewage lagoons (or waste stabilization ponds) in energy-impacted communities at least as a temporary measure to expand existing treatment capacity until more conventional secondary treatment facilities can be constructed. The main advantages of this approach are that capital and operating costs would be substantially lower and construction lead-time would be shortened. Thus, sewage lagoons provide a flexible, economical means of providing municipal wastewater treatment capacity until the population stabilizes and the necessary front-end financing from the expanded tax base is available. The major disadvantage is the inability to consistently meet EPA's secondary treatment standards. While this problem could be dealt with by complete retention of wastes, this would also increase consumptive water use.

As discussed in Chapter 3, salinity problems are already an important issue in the CRB and for some streams in the Yellowstone River Basin. Although the primary sources of salinity are agriculture and natural salt flows, energy development will increase the problem through salt loading and the downstream concentrating effects of consumptive use. Two policy options specifically addressing salinity issues are discussed. Constructing a range of salinity control projects is the policy approach currently being emphasized in the CRB. The other choice, a salinity offset policy, is defined specifically to deal with salinity increases caused by energy development. Controlling salinity has both environmental and economic benefits. The economic benefits for agriculture, municipalities, and industry for each mg/l decrease are estimated at $343,000 per year.

Because constructing salinity control projects is a more comprehensive alternative, it can have a much greater effect on salinity levels. For example, three projects authorized by the CRB Salinity Control Act (Grand Valley, Paradox Valley, and Las Vegas Wash) are estimated to decrease salinity concentrations at Imperial Dam by about 43, 18, and 9 mg/l, respectively. The disadvantages are that large federal capital outlays are required and water availability is decreased. Cost estimates for the three projects identified above range between $0.2 and $1.0 million per mg/l salinity reduction at Imperial Dam. Although there is some opposition to this approach and the authorized projects have been subject to considerable delays, the legal and institutional basis already exists through the CRB Salinity Control Act and the CRB Salinity Control Forum.

In comparison, the salinity offset policy is not intended to address salinity problems generally, but rather establishes an administrative mechanism to prevent increased salinity levels due to energy development. The primary advantage is that this approach should be economically efficient, since energy developers could use the lowest cost "offsets" available. Increased energy costs are expected to be small; for example, if offsets were obtained simply by paying a proportion of the costs for the salinity control projects discussed above, energy costs for three facilities analyzed were estimated to increase only by 0.02 to 0.5 percent. There are two important disadvantages to this policy. First, the salinity models required to predict salinity impacts are not well developed. Second, by adding another requirement to energy facility siting permits, energy development could be slowed or constrained. However, states could help to alleviate this potential problem by, for example, the banking of offsets. The legal basis and practical feasibility of implementing such a policy is uncertain. One type of salinity offset policy has been attempted in Colorado by a local government agency but the plan was vetoed by the governor. This plan would have affected a number of future water users in the state and was therefore strongly opposed by several groups. By limiting the offset policy to energy development activities the opposition would be reduced, although important equity questions would remain.

NOTES

[1]Recently, the EPA established the National Center for Groundwater Research, a research group operated by a consortium of three universities: the University of Oklahoma, Oklahoma State University, and Rice University.

7
State Water Administration and Management

Pressures for Improving Water Management

Western state governments have traditionally played the primary role in making policy for water resources. However, several changes have occurred in recent years which have increased pressures for states to improve their capacity to manage water resources. Table 7-1 summarizes the problems and issues identified in this report which affect state water administration and management. Although these problems and issues reflect a broad range of complex questions, they present three specific problems for state governments: (1) continually increasing demands for scarce water resources, including requests for many uses which represent relatively new demands on the system, such as energy and environmental needs; (2) increasing recognition of the relationship between water quantity and quality; and (3) allegations by some federal government officials, environmental interests, and others that state water policies encourage wasteful and inefficient use of the resource, and the corresponding threat of an increased federal role in western water management. Because of these problems, most western states increasingly recognize their own self-interest in improving their capacity to manage water resources.

This chapter is primarily concerned with two central activities of administering and managing water resources: managing water on a day-to-day basis to make it available, monitor its use, and protect its quality; and planning orderly change in response to new developments in energy, agricultural, municipal, and other activities. The remainder of this section describes existing state water management and administrative systems and identifies alternatives for improving these capabilities. This includes an overview of the major elements of state water management and administration, criteria

199

TABLE 7-1: Summary of State Problems and Issues

Problem Area	Summary	Chapter and Section
Competition among major sectors	Agricultural, energy, other industrial, and municipal users are making increasing demands for a scarce resource.	2.5
Environmental needs	Instream flow requirements are becoming increasingly important in several states to protect public health, fishing, and scenic values.	2.5.2
Reserved water rights	Federal reserved rights commit unknown quantities of water to federal and Indian uses which may reduce quantities available to current users.	2.6
Jurisdictional disputes	Conflicts among the states and between states and the federal government over questions of authority and responsibility for water management increase the uncertainty and complexity of water policymaking.	2.7
Water pollution from energy development	Liquid effluents discharged into holding ponds and solid wastes can pollute surface and groundwater; mining and in situ oil shale and uranium recovery can contaminate aquifers.	3.2
Salinity control	Agricultural, environmental, industrial, and municipal damage is already caused by salinity. Salinity problems can be increased by energy development and could soon affect the availability of water in the CRB.	3.3
Municipal wastewater pollution	The rapid population influxes associated with energy development can add to water pollution problems due to inadequate treatment of municipal sewage.	3.4

for evaluating changes in these systems, and a selected set of specific policy alternatives. The remaining sections of this chapter discuss, evaluate, and compare four policy alternatives;

- Reduction of institutional constraints to water transfers;

- Water rights exchange systems;

- Time-limited permit systems; and

- Coordinated administrative structures.

State Administrative Systems

As discussed throughout this book, the water policy system includes a complex set of international treaties; federal and state laws, regulations, and court decisions; and interstate agreements. Within this framework, management of water resources within state boundaries can be characterized by three broad activities: appropriation, adjudication, and distribution. The cornerstone of this system is the appropriation function which is carried out by state water engineers, water courts, and/ or water boards. The general appropriation responsibility of state governments is to determine if unappropriated water exists in a source to meet a given application. If the water does exist, a water right can be granted, subject to the provisions of the beneficial use, nonimpairment, and other restrictions which have developed in western state water law (see Chapter 2). When a right is granted, states record the holder, date, seniority, and specific use to which it will be applied.

This information is critical to the second major administrative function of state governments, adjudication. This function is intended to protect right holders from impairment or loss of water use. Adjudication can be initiated by a single user, for example, if a right holder believes the right is being illegally impaired by another user, or by a state in order to make a general determination of rights from a water source. For example, a general adjudication was initiated by New Mexico in 1975. The state filed a suit on behalf of San Juan County to determine all rights to the use of the surface and groundwaters of the San Juan stream system.

The third major function, distribution, is largely the responsibility of private sector groups (such as ditch companies), municipalities, and quasi-public agencies (such as conservancy districts), rather than state governments. Once a right is granted, for example, to a ditch company, state administration and management of the resource is ended except in the cases of application for transfer of rights or adjudication of the legal right to water. Ditch companies or conservancy districts then may sell and distribute water to individual users. However, the kind of use to which it is put, such as agricultural or industrial, is specified at the time of the grant application.

Three points are important regarding state water administration. Most importantly, states primarily play a bookkeeping role to ensure that water is used legally rather than a role which would monitor and manage how water actually is used. As stated by Harris Sherman (1977, pp. 226-27), former director of the Colorado Department of Natural Resources:

> ...up to now the State has played a ministerial role as opposed to a managerial role in water. We have been the bookkeeper and the referee. The major function of both the State Engineer and the courts has been to keep the books and to be the referee of water use by private parties. The main emphasis of state involvement has been to facilitate the use and to provide enforcement for water users.

Secondly, state water administrative systems are subject to change, as evidenced by recent inclusions of instream values as a beneficial use in several western states. However, broad or rapid change within these systems is extremely difficult. This is in part due to the international, national, and interstate influences on western water resources. It can also be attributed to the long history of water use in the West, in which water was valued primarily for agricultural and munici- pal uses. Difficulty of change is also due to an overreliance on the courts as a problem-solver in appro- priation systems. As this developed over decades, interpretation of the legality of water use has become exceedingly complex. In addition, the courts have be- come the major institution of change, yet they operate very slowly and create piecemeal, localized and short- term resolution to problems.

Thirdly, substantial variation exists among the eight states regarding water policy and administration. For example, beneficial use is defined differently in each state, authorization for transfer of water varies among water courts and state engineers, and the length of time required before rights are forfeited by aban- donment varies from three years to ten years (Radosevich

1978; White 1975). Variation also exists in the values and interests concerning how water should be used. For example, among the eight states, Utah is one of the strongest in favor of water for energy development; in contrast Montana has been active in protecting the environmental benefits of water.

Criteria for Evaluating Management Alternatives

A variety of criteria can be used to evaluate the water management policy options considered in this chapter. The following criteria reflect the most pressing needs which states have for dealing with the water problems and issues.

- Better Information: Will the alternative improve the information base about current use patterns?

- Flexibility: Will the alternative help states to meet multiple demands for water?

- Integration: Will the alternative improve states' ability to consider and administer water quantity and water quality together rather than separately?

- Economic Efficiency: Will the alternative allow water to be transferred to uses with a higher economic value?

- Social and Environmental Effects: Will the alternative minimize social and environmental costs and risks and maximize the benefits derived?

- Feasibility: Will it be possible to adopt and implement the alternative?

Policy Alternatives

Two general categories of alternatives will be considered in the remainder of this chapter: alternatives to facilitate water transfers and alternatives to improve management and planning of water resources. As shown in Table 7-2, several specific alternatives will be evaluated within these two general categories. Facilitating water transfers

TABLE 7-2:   State Management Alternatives

Category	Specific Alternative
Facilitate water transfers	Reduce institutional constraints to transfers
	Modify beneficial use requirements
	Modify nonimpairment provisions
	Reduce appurtenance requirements
	Local control of federal projects
	Develop water rights exchange systems
	Develop time-limited permit systems
Improve management and planning	Coordinated administrative structures

can include reducing institutional barriers, developing mechanisms for water rights exchanges, and making permits time-limited. Coordinated administrative systems will be discussed as an option for improving management. These alternatives do not represent the full range of administrative, management, and planning choices available to states. However, they do represent several options which states have either considered or attempted, and they are responsive to the water problems and issues likely to face the states over the next several decades.

REDUCING INSTITUTIONAL CONSTRAINTS TO TRANSFERS

Existing laws and regulations of the appropriation system and federal reclamation projects are a two-edged sword; they provide the legal framework for protecting individual water rights, but they also create barriers to transfers of water rights to meet changing needs. Among the most important barriers are: (1) beneficial use, which can restrict transfers from one kind of use to another; (2) nonimpairment, which protects third parties affected by transfers; (3) appurtenance, which legally attaches water to land; and (4) federal project controls, which create an ambiguous state-federal relationship regarding authority over water allocation from federal projects. The following sections discuss how each of these barriers may be reduced.

Beneficial Use and Public Interest

Perhaps the biggest constraint on water right transfers is the beneficial use doctrine (see Chapter 2). This doctrine, originally intended to limit water waste, has also been used to restrict certain uses of water and water right transfers. The Montana legislature, for example, determined that "a use of water for slurry to export coal from Montana is not a beneficial use" (Mont. Code Ann. 1978a). Water transfers from agriculture to industry are also prohibited in Montana if they involve water flows of 15 cubic feet per second or approximately 11,000 acre-feet per year (AFY) (Mont. Code Ann. 1978b). Even greater restrictions exist on transfer of water out-of-state. In Wyoming, for example, the appropriation, diversion, or storage of the state's surface or groundwater outside the state is prohibited without the approval of the legislature (Wyo. Stat. Ann. §41-3-105). These restrictions could be reduced by declaring water rights transfers themselves a beneficial use. Idaho has done this to implement its water supply bank (Idaho Code 1979a).

Nonimpairment

Another limitation on water rights transfers is satisfying the nonimpairment doctrine, which generally prohibits transfers which would injure the rights of third parties. As described in Chapter 2, resolving nonimpairment questions in court is an expensive and time-consuming process. This barrier could be reduced by establishing administrative procedures to settle disputes over impairment or to compensate third parties. The State of Idaho permits third parties to appeal for modification of the quantity of water transferred in some transactions. If right holders claim impairment, they may appeal to the Department of Water Resources. Upon investigation, the state may amend or revoke the transaction if third party rights are, in fact, harmed. Further appeal by any party could then proceed through the district courts (Idaho Code 1979a).

The constraint of nonimpairment could also be reduced if adequate data were available to determine consumptive use. If users were selling only consumptive use, the rights of downstream users would not be impaired. However, the technical difficulties of quantifying consumptive use for individual users are severe.

For example, if an irrigator withdrew 3,000 AFY, and
2,000 AFY directly returned downstream, it is not accurate to assume that 1,000 AF was consumed. Actual consumptive use quantification would require knowledge of
the quantity of water consumed by the plants irrigated
and the amount which is irretrievably lost, such as
through percolation to deep aquifers. Thus, reducing
the institutional and legal problem surrounding water
rights quantification is an important alternative for
reducing nonimpairment constraints. This quantification
and information problem is discussed further in Chapter
8.

Appurtenance Requirements

For most states in the study area, water rights are
not separable from the land, except if it becomes impractical to irrigate beneficially. Five states (Arizona,
Montana, New Mexico, North Dakota, Wyoming) require judicial or administrative review of requests for separating water rights from land. Colorado and Utah are the
exceptions to this situation; in Colorado, for example,
any part of a water right may be retained or conveyed
with the land, depending on the specifics of the contract (Nielson v. Newmeyer 1951).

Water delivered to lands in projects of the Water
and Power Resources Service (WPRS) (formerly the Bureau
of Reclamation [BuRec]) is typically considered appurtenant to land irrigated (U.S.C. 1976; and Reclamation
Act 1902, §8). Appurtenancy is generally enforced by
prohibiting irrigation districts from transferring water
outside the district as a condition of the water supply
contracts with WPRS. The Southeastern Colorado Water
Conservancy District, for example, is prohibited by its
contract with WPRS from transferring water outside the
district, except to municipalities (Hartman and Seastone
1970). Water transfers between users in one district
are possible, but this must be specified in the terms of
the contract.

As this discussion suggests, appurtenancy does not
prohibit water transfer. It can make it more difficult
by requiring a user who wants to transfer a right to go
through the necessary administrative or judicial process, including, in most cases, proving that other
rights would not be impaired.

Water Right Transfers from Federal Projects

A fourth constraint is the ambiguous relationship
between the WPRS and states. The National Water Commission has suggested that the law is far from clear

regarding the WPRS's title to water, the nature of the right, and its powers regarding transferability (U.S., NWC 1973, p. 264).

Confusion regarding jurisdiction over water allocation from federal projects increased in 1975 with a Memorandum of Understanding signed by the Secretary of the Army and the Secretary of the Interior, giving Interior the authority to market water to industrial users in the Missouri River Basin (MRB).[1] WPRS offered "water contracts" to the ten states of the MRB, who could then market the water to industrial users. Both the federal and the respective state governments had claims for ownership of water in the six reservoirs, but the marketing program was not based on a claim of ownership (U.S., DOI, BuRec, Upper Mo. Reg. 1976). The marketing program required "a repayment for the cost of the service provided by storage and regulation of the waters, which would otherwise not be available without the main stem storage system" (Humm and Selig 1979; and U.S., DOI, BuRec, Upper Mo. Reg. 1976). North Dakota refused the water contract offer, in part because acceptance would legitimize the authority of the WPRS to allocate water in the MRB (Humm and Selig 1979).

The relationship between the federal and state water agencies has been clarified somewhat in California v. United States (1978). In this case, the Supreme Court held that the federal government must observe state procedures and conditions in regard to obtaining water rights, unless federal law provides otherwise.

Instead of attempting to unravel the complex relationship between states and WPRS, an alternative is to allow states to control the transfer of water from federal projects. This could allow transfers of both surplus water and water currently being consumed by particular users, such as agriculture. Transfers would be subject only to state laws and regulations, eliminating the approval requirement by the federal water-supply agency (U.S., NWC 1973, p. 266).

Evaluation

Table 7-3 summarizes the advantages and disadvantages of these alternatives. Reducing constraints to transfers and thereby increasing the flexibility of state water systems is likely to become increasingly important as agricultural, municipal, energy, environmental, and other users compete for a scarce resource. These alternatives are also of value in improving the economic efficiency of water use. This would occur because the costs of transfers would be lower and because transfers from lower-valued to higher-valued uses would be encouraged. The costs of transfers would be lowered

TABLE 7-3:   Summary of Alternatives to Reduce
             Institutional Constraints

Criteria	Findings
Improved information	These alternatives will gradually improve information on water resources if consumptive use is quantified as part of the transfer process.
Flexibility to meet multiple use	Barriers to water transfer, particularly procedural requirements, are reduced.
Integration of quality and quantity	Comprehensive or integrated management is not improved by these choices.
Economic efficiency	Economic efficiency is encouraged as water is allowed to be transferred to industrial users. Individual farmers will benefit financially.
Social/environmental effects	The agricultural sector will be hurt over the long term.
Feasibility	These choices are gradually being introduced in the West. However, continued resistance from some agricultural interests can be anticipated.

by eliminating appurtenancy requirements, thus reducing the time and expense of legally separating water rights from land. Costs could also be lowered by establishing procedures to satisfy the nonimpairment doctrine. One study estimates that the engineering and legal costs associated with a contested transfer in the San Juan basin average about $4.36 per AFY (Khoshakhlagh 1977). Quantifying water rights on the basis of historic consumptive use would reduce the expense of litigation to prove nonimpairment, but this reduction may be offset by the costs of data collection.

These alternatives would also have the advantage of increasing the flexibility of the water policy system by increasing the opportunities for multiple uses of water. They would also help state governments gain more control over federal project water. In addition, quantification of consumptive use would improve the data base on use patterns, thereby enhancing day-to-day management capacity.

Although transfers are encouraged by these alternatives, several trade-offs will be apparent. Social effects are a concern since these alternatives threaten the dominant position of agriculture in the western economy. Water would probably be transferred from agricultural uses to the municipal and energy sectors, leaving individual farmers and companies better off financially, but reducing farm-related income and employment for the agriculture sector as a whole. Opposition from agricultural interests decreases the feasibility of these choices. Feasibility is further decreased by the probability that some state agencies, such as water management boards, and the WPRS would be losing some control over water allocation.

WATER RIGHTS EXCHANGE SYSTEMS

Water rights exchange systems, such as a water bank, would establish an explicit mechanism for transferring water rights. Such mechanisms could be established within states or on a regional level. The following discussion describes water rights exchange banks and interbank transfers, provides an example of water bank management, and evaluates this alternative for facilitating water transfer among sectors.

Water Rights Exchange Banks and Interbank Transfers

The purpose of water rights exchange banks is to coordinate demand for water rights with water availability. Like financial banks, they would offer interest payments (a rate of return) for a depositor's rights and allow a borrower to make withdrawals upon payment. Deposits and withdrawals would consist of the rights to the water, but not necessarily the water itself. That is, depositing water rights in a water bank does not involve physical movement of water. For example, an irrigator who deposits rights to 1,000 AFY for a specific period of time would not be required to forego consumption of the water entitled by those rights until a buyer actually wanted the water for use.

Buyers would provide a mechanism for transporting the water. Thus, most transfers would be limited to the existing network of water transportation mechanisms (canals, ditches, pipelines, and storage reservoirs) unless buyers could afford their own transportation facilities. Energy companies generally can better afford the cost of transporting water than can agriculture.

Water rights exchange banks could be established privately, as a cooperative enterprise among irrigation districts, or publicly, as part of state water regulatory agencies. The water supply bank of Idaho, for example, is the responsibility of the Department of Water Resources (Idaho Code 1979b). In many areas, banks could benefit from state cooperation with the WPRS to take advantage of existing transport facilities.

Pricing of water rights would be competitive under this system, reflecting both the scarcity of water, and the value of the right in terms of its seniority, quality, and the length of time required for transfer. For example, a water right with an early priority date would likely command a higher price than a water right of an equivalent quantity and quality but with a later priority date. Rights for saline water would probably bring a lower price than those for fresh water. Also, water rights sold outright or leased for relatively long periods of time would bring a higher price for the depositor than those leased for shorter periods.

Idaho's Department of Water Resources established a water rights exchange bank similar to that described above. Water rights are deposited in a bank by filing an application indicating: (1) that the water right had been recorded; (2) proof of ownership and that the right had not been forfeited or abandoned; (3) evidence that water was available to fulfill the right; and (4) evidence of concurrence by an irrigation district, if necessary (Idaho, Dept. of Water Resources 1980). Withdrawals from the bank are accomplished through periodic announcements of the availability of water for purchase or lease from the bank, specifying a deadline for accepting applications. The director of the Water Resources Board has the power to approve or reject withdrawal applications that request leases of less than two years' duration. Applications for longer than two years or purchases of water rights must be approved by the Water Resources Board (Idaho, Dept. of Water Resources 1980, p. 5).

Although more difficult from an engineering standpoint, interbank transfers would facilitate water reallocation at a regional or basin-wide level. Figure 7-1 suggests an interbank transfer mechanism. Water right transfers between local banks would be coordinated through a regional bank. The supply and demand of water rights at regional bank levels could be coordinated at the subbasin level such as the Upper Colorado River Basin. Information regarding quantity of water rights, demand, prices, quality, and seniority could be transmitted vertically and horizontally through the system of banks thus increasing opportunities for water rights transfers over large areas.

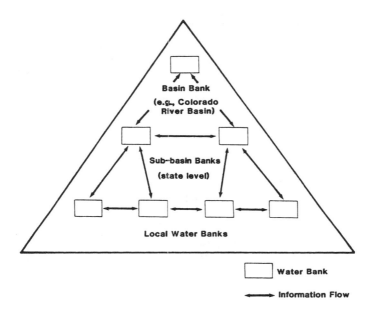

Figure 7-1:   Hypothetical Water Bank System

An Example of Water Bank Management

In this section, a hypothetical water bank scenario
is constructed for the oil shale area of northwest Colorado in order to describe how such a system might respond to energy-related water needs.   The hypothetical
examples described below can be contrasted with the approach currently pursued by Colony Development Corporation, which intends to purchase water from upstream
reservoirs and transport it to the production area in
the Piceance Creek Basin.   Colony initiated negotiations
with WPRS for annual delivery of up to 7,200 acre-feet
(AF) of water from the 152,000 AF of storage in Green
Mountain Reservoir.   The contract would span a period of
twenty years, with a renewal possible (U.S., DOI, BLM
1977).   The proposed sale to Colony amounts to a twenty-
year lease of the water rights.

Under a water bank system, WPRS would deposit the rights to 7,200 AFY of water in Green Mountain Reservoir, which would serve as a state-administered water bank. The deposit in this case would be limited to twenty years. The state would pay WPRS a rate of return for its deposit based on the relative seniority of the water right, the length of the deposit time, and also the expected demand for the right. For example, if demand is relatively low, depositors would receive a lower rate of return than when demand is relatively high. Colony would pay the state for its withdrawal. The value of this system would be to eliminate the lengthy federal and state review and judicial process. Challenges to decisions would be permitted under civil court procedures.

Water bank transactions among several users would be more complex. Duration of deposit could vary from one irrigation season to "in-perpetuity." The water bank would inform interested consumers of water availability, price, the period of time of availability, and seniority. If the prices, quantities, and time periods of availability are agreeable to the buyer, and if the transfer would not harm third parties, the transfers would be activated on the date specified by the withdrawing party.

Figure 7-2 shows the hypothetical case of two consumers and three depositors. Physical changes in flow are made by changing withdrawals and releases. As indicated in Table 7-4, these bank users might have water rights deposits of 65,000 AF. Depositors must reduce their withdrawal of water by specific quantities and periods of time when notified by the water bank of the activated transaction. But until the transactions are activated, that is, until Colony and user D can put to use their 21,000 AF of water represented in the water rights transaction, the depositors may continue using the water. Water rights deposits are classified as either sales or leases. Hypothetical rate of return, price paid to the depositor, and the time period of the deposit are shown. The last set of entries under "deposits" shows activated transfers. That is, depositors must reduce withdrawals by the quantities and lengths of time specified by the water bank.[2]

The hypothetical water rights exchange bank would also provide coordination for water rights transfers for nonconsumptive uses. For example, Colony could lease 2,000 AFY from the water bank in order to ensure minimum stream flow during the irrigation season. Water rights for nonconsumptive use could also include a salinity offset policy (see Chapter 6). A salinity offset policy would require water quality improvements if consumptive use of low salinity water reduces downstream dilution. If salinity increases from water withdrawals, a company

213

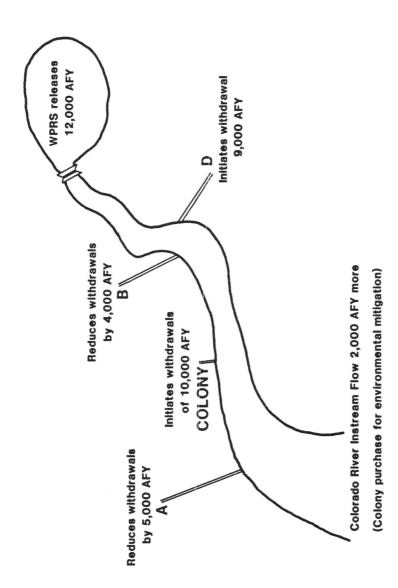

Figure 7-2: Net Physical Water Transfers in Oil Shale Region of Northwest Colorado (assuming water rights transfers are "activated").

TABLE 7-4: Water Rights Exchange Bank Accounting (hypothetical)[a]

Transactions	Colony		Transaction 1	Transaction 2	WPRS	D	Total AFY
	A	B					
Deposits							
Permanent acquisition (AFY)	4,000	7,000					11,000
Seniority[b] (year)	(1940)	(1945)					
Price ($/AFY)	$5,000	$4,000					
Time deposit (AFY)	6,000	8,000			40,000		54,000
Seniority (year)	(1940)	(1945)			(1935)		
Return on deposit ($/AF)[c]	$50	$40			$80		
Time period	(1979)	(1979-1985)			20 years		
Total bank deposits (AFY)	10,000	15,000			40,000		65,000
Activation to meet withdrawal (period of activation)	4,000 (perpetuity) 1,000 (1979)	4,000 (1979-1985)			2,000 (1979) 10,000 (1979-1999)		
Total activated	5,000	4,000			12,000		21,000
Withdrawals							
Sales (AFY)						4,000	
Seniority (year)						(1940)	
Price ($/AFY)						$110	
Leases (AFY)			2,000	10,000		5,000	
Seniority (year)			(1935)	(1935)		(1940 & 1945)	
Price ($/AFY)			$90	$100		$90	
Time period of lease (years)			(1979)	20 years			
Date of activation			June 1, 1979	June 1, 1979		yearly June 1, 1979	
Total bank withdrawals (AFY)			2,000	10,000		9,000	21,000

$/AFY = dollars per acre-foot per year

[a]Quantity defined by historic amount of consumptive use.

[b]Original date of filing for water rights.

[c]Rate of return paid by depositors.

such as Colony would have to find some method of offsetting the increased salinity effects of a proposed
water transfer. If water banks were established, Colony
could lease or buy water rights to offset salinity increases from its withdrawal upstream. The water obtained by the activated water rights transfer would be
allowed to flow downstream, offsetting the increased
salinity effects of the upstream withdrawal.

Evaluation

It is difficult to evaluate water rights banking
since little experience exists. One recent experience
was the Emergency Drought Relief Act of 1977 which authorized the secretary of the interior to purchase water
from willing sellers for distribution to farmers with
highly-valued crops. BuRec purchased 211,435 AF of
water, including 20,250 AF from the Metropolitan Water
District of Southern California. This water was distributed through existing transport facilities to user
organizatons in reclamation projects. Although the
water was relatively expensive, the adverse effects of
the drought were somewhat mitigated (U.S., DOI, BuRec
1978c, p. 43).
An evaluation by the General Accounting Office (GAO)
found that of the $75 million authorized by Congress for
water banking operations under the Act, only $4.8 mil-
lion was obligated to purchase water from sellers. The
GAO concluded that the program was largely ineffective
in most of the western states, except California, where
it was moderately successful. A large part of the inef-
fectiveness was blamed on the timing of the Act. Most
farmers had committed their water supplies to a definite
use by the time the Act was implemented, thus restricting the quantity of water available for banking purposes. Nonetheless, the water banking concept has been
shown as an effective method of reducing drought impacts
in California (U.S., GAO 1979).
In addition to dealing with drought situations,
water banks have several potential applications for the
eight-state study area. Table 7-5 summarizes the advantages and disadvantages of this alternative. Its greatest strength appears to be its flexibility. If potential
new water users can acquire rights through sale or
lease, scarce water can be distributed according to demand among agricultural, municipal, and industrial sectors of the West. This would be particularly advantageous for energy developers in acquiring senior rights
which would be subject to less risk of being unavailable
during droughts.

TABLE 7-5: Summary of Water Rights Exchange Systems

Criteria	Findings
Improved information	Water rights exchange systems would increase knowledge about water availability and demands.
Flexibility to meet multiple uses	The primary advantage of this alternative is in facilitating transfers to make water available for more uses.
Integration of quantity and quality	Opportunities would be increased for integrating quality and quantity, for example, by facilitating salinity offset policies.
Economic efficiency	Transfers to economically higher valued users would be encouraged. Speculation would be encouraged.
Social/environmental effects	This alternative would probably result in loss of agricultural lands and changes in the economic structure of the West. Energy development would be enhanced.
Feasibility	Changes would be required in the beneficial use doctrine and appurtenance requirements. Opposition from some agricultural interests would be expected.

A constraining factor of water banks is that actual transfer of water, contrasted to transfer of water rights, is limited by the availability of water transporting facilities, such as ditches, reservoirs, canals, and pipelines. A successful example of water banking existed during the 1976-77 drought in California, but this is a state with a highly developed network of water pipelines, ditches, and canals (U.S., GAO 1979). While some users may be willing to invest in pipelines or ditches, the added cost of the transportation system may not be acceptable unless rights are available over a long period. Long-term leases may bring higher water prices; this would be to the advantage of energy industries which are more likely and more willing to pay the necessary price to obtain a long-term supply.

The information-gathering function of water rights exchange banks will provide data concerning water quality, price, and availability for local areas. Regional water banks would facilitate data gathering for a larger area. This characteristic of water banks would help to reduce one of the largest constraints on meeting multiple water demands--locating water rights which are for sale and determining a fair price for those rights. In essence, this alternative could help to establish an organized brokerage industry for water rights.

Water banking could also be useful to protect water quality and improve the integration of water quality and quantity. For example, salinity offset policies could be established concurrently with water banks to preserve or improve water quality. One way of reducing the concentrating effects of withdrawals on a watercourse, for instance, is to lease enough agricultural water rights to offset the salinity concentrating effects of the proposed withdrawal. Water banking would permit leasing of those rights, thus providing an alternative to water rights purchases from agriculture. Leases would permit transfers for a limited amount of time, while purchases would in effect transfer water rights from agriculture indefinitely.

As noted in the previous section, any improvements in the water rights transfer process may allow water to be obtained for energy production to the detriment of irrigation. Under the market pricing scheme outlined above, agricultural users would generally not be able to compete with industry or municipalities for extra water in periods of drought when water right prices would be high. According to its advocates, market water rights pricing will encourage economically efficient use of water, by allowing the price of water rights to reflect the scarcity of water (Davis and Hanke 1971). Water rights would then be allocated among users according to the value of their output per unit of water input. The result would be the transfer of water rights from low-value uses to high-value uses. However, a more thorough investigation of the impact of water allocation under competitive bidding needs to be undertaken in order to accurately assess the economic efficiency of market water rights allocation.

A competitive bidding system for water rights would change the distribution of economic costs and benefits among different types of water users and economic sectors. This change would be primarily due to new prices for water rights, but the actual price changes are difficult to predict. The water rights exchange bank would enhance the availability of rights through an organized marketplace, and prices might fall. More likely, however, an upwards bidding pattern for a scarce resource would result. The increased cost would be a small fraction of the total costs for expanding industries and municipalities, but a large proportion for agriculture. Increased prices would benefit existing water rights holders, primarily water districts or landowners.

Higher prices for water rights may also reduce the amount of land in agricultural production. In this case, workers who depend on agricultural production and businesses which depend on income from selling farm implements and seed would be adversely affected. Reduction in farm employment would also adversely affect local agricultural economies. Labor displaced from farming may not be properly skilled to fill positions created by the expanding energy sector.

TIME-LIMITED PERMITS

Under the appropriation system, water rights are granted in perpetuity, similar to a property right. Because the transfer of water is inhibited by nonimpairment, appurtenancy, and other aspects of the appropriation systems, it may be in states' interest to limit new permits to specified time periods. This alternative would allow water to revert to the state when the specified time period ends.

Inclusive vs. Selective Permits

Inclusive time-limited permits would be intended for all new requests for unappropriated water. This idea is modeled after the approach used in Florida and Iowa. In Florida, permits for consumptive water use are limited to a maximum of twenty years. To obtain a permit, the request must be for a reasonable beneficial use, not interfere with any present legal use, and be consistent with the State Water Plan (Fla. Stat., §§371.06 through 373.201). An inclusive permit system has also been established in Iowa, where permits are required for all but domestic uses, with a maximum duration of ten years. No priority system exists and the Water Commissioner has the authority to suspend permits during a drought. Thus, the Iowa state water agency has more control over water distribution than in appropriation states (Maloney, Plager, and Baldwin 1968).

Under the inclusive permit system, the priority of agricultural, environmental, municipal, and industrial permits and the amount of water they would get during a water shortage could be specified. If a state placed a high value on industrial development, for example, it could protect that use of water during shortage.

Selective time-limited permits would be applied to particular sectors of the economy such as industry, mining, manufacturing, or other relatively time-limited activities. Once the permit expired, the water would revert to the state. Such an industrial time-limited permit system is used in Utah (Utah Code Ann. §73-3-8).

Selective permitting applied to industry would aid state water management in two ways. First, the state would gain more control over the use of water by a changing industrial base. This would limit the permanent removal of water from agricultural use. Secondly, as the time-limited permits expired, the state would have a source of essentially new water to appropriate. This water could be very important by the year 2000, when some states will have little or no unappropriated water in some streams.

Evaluation

Table 7-6 summarizes the advantages and disadvantages of inclusive and selective time-limited permits. The primary strength of these alternatives is that they increase states' flexibility in meeting changing demands for water. Although agricultural interests may resist any attempt to issue time-limited permits or otherwise change the current system for regulating water use, this alternative appears less threatening to agriculture than many policy options. In general, the inclusive time-limited permit system could not be imposed on existing water right holders without statutory changes (Maloney and Ausness 1971, p. 527). Even if applied only to new water rights, other features of western water law could make this alternative difficult to apply to agriculture. The most significant of these is the nonimpairment doctrine.

For these reasons, a time-limited permit system applied selectively to industry or to critically water-short areas might be more acceptable. Agriculture would be less threatened with an industry-only permit system, and in fact, could benefit since the industry time-limited permits would prevent energy developers from securing water in perpetuity. Industry's response would probably depend on the duration and priority of the permit. Industry would have to be assured that the duration of the permit would be sufficient to at least recover investment costs.

The selective time-limited permit system should reduce conflict between energy developers and agriculture by assuring agriculture that water used for industrial purposes would revert to the states for reappropriation after completion of the industrial activity. In this sense, the selective time-limited permit system should

TABLE 7-6:   Summary of Time-Limited Permits

Criteria	Findings
Improved information	Does not directly improve the existing information base.
Flexibility to meet multiple uses	Can increase state flexibility over the long-term by allowing for reevaluation of water uses and needs.
Integration of quantity and quality	Does not directly improve integration.
Economic efficiency	Encourages economic efficiency by increasing likelihood of water use for highest-valued uses.
Social/environmental effects	Although limited to streams/rivers with unappropriated water, valuable for a wide range of social purposes because of flexibility and reversibility.
Feasibility	Less threatening to agricultural interests than many alternatives. Requires statutory or constitutional changes in most states.

increase the acceptability of energy development in the West.

COORDINATED ADMINISTRATIVE STRUCTURES

Each of the three previous alternatives is intended to increase state government control of water resources primarily through changes in the legal and regulatory doctrines of the appropriation system. Each of these alternatives would be enhanced by a coordinated administrative structure which would increase responsibility for day-to-day management of water and would explicitly integrate water quality and water quantity concerns. A coordinated administrative structure could take a variety of forms, ranging from a centralized Department of Water Resources to a "bottom-up" process which would delegate significant planning authority to local areas. This section discusses an administrative system which would combine elements of "bottom-up" planning with centralized agency control.

The Utah Plan

In 1975, Dewsnup and Jensen developed a water administration plan sometimes referred to as the "Utah Plan" to provide for local participation in water resources planning within the context of an integrated and centralized state administrative system. Although never implemented, the basic idea of this plan, as shown in Figure 7-3, is to organize water resource planning around local areas, or hydrological units. Such units could consist of part of a river basin; for example, they could coincide with regional 208 districts. Within each unit, water planning staffs, in cooperation with local interests, would collect data and formulate the basics of a tentative water plan. This plan would identify the current water supplies, usage, and alternatives for conserving and augmenting that supply. Augmentation methods might include cloud seeding, water savings practices, groundwater development, conjunctive management of surface and groundwater supplies, and reclamation of brackish water, among others (Dewsnup and Jensen 1975, vol. 1, p. 2-22). The water quality impacts of the plan would be assessed through data collection efforts coordinated with water quality planners.

Tentative district plans would be submitted to a Unit Advisory Council composed of various interest group representatives concerned with water management within that district. These interest groups would include local, state, and federal government representatives, and representatives of environmental groups and industry. The Unit Advisory Council would review the tentative district plan and identify preference and priorities concerning augmentation methods, water conservation practices, and types of development within the district.

When this plan was proposed for the State of Utah, in addition to the Advisory Council review, the district plan was also to be reviewed by representatives of relevant state agencies. These included members of the State Advisory Planning Committee, Division of Water Resources, and the State Engineer. Specific water quality issues could be reviewed by the Department of Health and other water quality agencies. In contrast with the Advisory Council, agencies represented on the committee would not have the authority to actually alter the draft plan for the district, but rather would simply review and make comments and recommendations. Comments, recommendations, and the draft plan would then be sent to the agency having overall authority over the state's water resources. In the case of Utah, this would have been the Board of Water Resources. For other states in our study area, such a plan would require a new and perhaps centralized water agency.

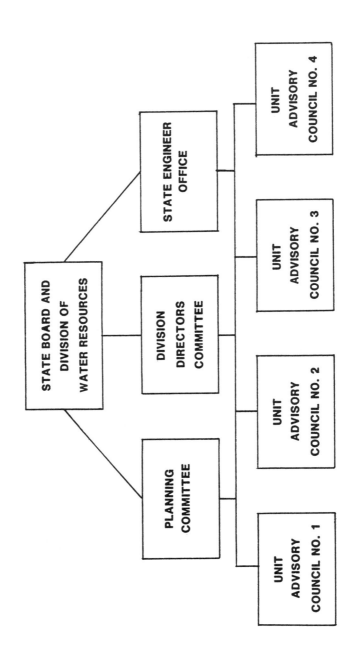

Figure 7-3: Institutional Arrangements for the "Utah Plan"

Source: Modified from Dewsnup and Jensen 1975, vol. 1, p. 2-28.

At this final step, the agency would hold public hearings on the plan. Taking the comments received from the hearings with the comments provided by the reviewing agencies, the Board of Water Resources or relevant primary agency would finalize and adopt the unit plan (Dewsnup and Jensen 1975, vol. 1, p. 2-25). There would be no judicial review of the agencies' decision since the plan per se does not impair or impinge on any person's property right. However, at a later point, if an individual felt that property interests were impaired by implementation of the plan, relief could be sought through the courts.

The "Utah Plan" for a decentralized district water planning system would serve as a legal guide to state and local decisions concerning water use, and would also serve to advise federal agencies. State and local water management agencies would be legally required to follow the unit water plans in allocating water. Departure from the plan would require demonstration that the public interest would be better fulfilled by not following the contested section of the unit plan. Unit plans would be reviewed and updated every year, incorporating recommendations submitted by the Unit Advisory Councils.

Evaluation

Table 7-7 summarizes the advantages and disadvantages of the "Utah Plan" as one example of coordinated water administration. The primary benefit of coordinated water administration would be increased capacity to integrate water quality and water quantity concerns. If this criterion is to be met, it will require new or substantially changed administrative systems throughout the study area. The second major benefit of the idea is flexibility; because diverse interests are represented at the level of hydrological units, multiple demands on water can be accommodated. For example, in Utah energy development could receive high priority in a unit representing the east-central part of the state while environmental or other values may take precedence in other units. This idea contrasts significantly with current state systems which typically place priorities for water use on a state-wide basis. Finally, the "Utah Plan" is explicitly designed to improve the quality of information about water resources by developing detailed information about hydrological units.

Such an approach could reduce much of the uncertainty faced by prospective users under the appropriation system. For example, a prospective appropriator would know from a unit plan if unappropriated water existed, if any existing rights were available for purchase and transfer, and if a specific use of water is

TABLE 7-7: Summary of Coordinated Administrative
Systems

Criteria	Findings
Improved information	An explicit purpose would be to develop information systems within hydrological units which would describe current use patterns, availability, quality, etc.
Flexibility to meet multiple uses	Flexibility is increased because each hydrological unit within the state could have different priorities for water use.
Integration of quality and quantity	A primary purpose is to integrate water quality and availability both at the local (hydrological unit) level and at the state administrative level.
Economic efficiency	Economic efficiency is not necessarily affected.
Social/environmental effects	Environmental concerns are made an explicit part of water planning.
Feasibility	The "Utah Plan" failed in Utah, in part because of political changes in the state and because of opposition by agricultural interests. It represents a significant departure from current administration practices.

compatible with the water use plan (Dewsnup and Jensen 1975, vol. 1, pp. 2-32, 2-33). This knowledge would be a considerable aid to the appropriate siting of energy facilities within a district or state, since the water available, the water quality requirements, and the local community's priorities and preferences would be known beforehand.

In most western states, current water administration is characterized by a statutory severance of the planning function from the management function. By combining the planning function of, for example, the board of water resources with the management regulatory powers already existing in many agencies (especially in the office of the state engineer), comprehensive water management planning would be facilitated.

The problems of pressure and resistance from local interest groups remain a serious impediment to establishing coordinated water administration. The "Utah Plan" directly challenged agricultural interests and the plan was rejected by the Utah legislature. Whether integrated planning and management would promote an acceptable allocation of water would largely depend on the degree of consensus among the various competing interests in a unit: agricultural producers, environmental groups, industrial development interests, energy producers, and Indian tribes. The failure of this idea to gain acceptance in Utah exemplifies the typical problems facing attempts to implement change towards more comprehensive water management planning. The current system seems quite adequate from the perspective of the traditional users such as irrigated agriculture who are generally satisfied with their allocations and subsidies (see Chapter 2). This position, however, may contrast with the position of some state-level decisionmakers interested in promoting municipal and industrial water uses which might yield greater opportunity for employment and income for the state's residents.

SUMMARY AND COMPARISON OF ALTERNATIVES

Four general alternatives have been presented for increasing states' capacity for managing water resources. These alternatives respond to two general situations affecting water resources: expanding demand for water to be used for multiple purposes; and increasing pressures to improve day-to-day management of water resources, particularly regarding the integration of water availability and water quality concerns. A summary comparison of these alternatives with respect to the six evaluative criteria used throughout this chapter is found in Table 7-8.

Each of the four alternatives is designed to improve flexibility, that is, to facilitate water use for a variety of purposes. While state administrative systems currently allow this to some extent, they reflect the primacy of agricultural and municipal water use. As other values continue to become more widespread, states will have several options for changing the legal, administrative, and regulatory system. The most important difference among the options is whether responsibility for meeting multiple uses will rest primarily with state governments or with the private sector. Coordinated administrative systems and public-operated water banks would provide the greatest degree of state control. In contrast, reducing institutional barriers to water transfers would gradually allow the market system to play a

226

TABLE 7-8: Comparison of State Water Management Alternatives

Alternatives	Current Status	Advantages	Disadvantages	Feasibility
Reduction of Institutional constraints to water transfers	Gradual changes are occurring in the appropriation system; four states have changed beneficial use definitions to account for instream values. However, states have made definitions more restrictive, for example, prohibiting coal slurries (Montana).	Flexibility of water management is increased by allowing multiple uses and economic competition for water. Some farmers will benefit substantially and, in general, industrial uses will increase. Gradually, the quality of information about water resources will be enhanced.	These options are very piecemeal approaches; comprehensive or integrated management is not directly enhanced. Over the long term the agricultural sector will be hurt as water is transferred to more efficient uses. Also, some changes will be made to limit rather than encourage multiple uses.	Among the choices considered in this chapter, this option is the most feasible since it represents incremental, gradual changes in water policy. However, it will be used sporadically due to opposition from agricultural interests.
Water rights exchange systems	Idaho established a water supply bank in 1979. A short-term banking system, established in 1977 by DOI in response to the drought, was used primarily in California.	The primary advantage of the system is in facilitating transfers to make water available to more users. This would promote more economic uses of the water. The system would also encourage improved information about water systems, and opportunities for integrating quality and quantity could be increased.	The system depends on the willingness of existing users to sell rights to the bank. It can also be constrained by the availability of water transporting facilities. For some users building new transportation facilities will increase total water costs. Although this option encourages multiple uses, it does not necessarily enhance state management capacity.	This option has been tried in the West and, in some respects, it is a formal system for what already occurs in many areas; that is, some farmers already rent their rights. However it would require changes in the beneficial use doctrine and appurtenance requirements and will be opposed by some agricultural interests.

TABLE 7-8: (Continued)

Time-limited permits	Time-limited permits are used in Florida and Iowa. In the eight-state study area, Utah has an industrial time-limited permit system.	The primary advantage of inclusive and selective permits is that they increase state flexibility in meeting new demands. This option could help to reduce conflict between industrial and agricultural users since industrial users could acquire rights which would ultimately revert to the state.	Permitting can only be applied to new water users under the appropriation system; thus, it is primarily intended for energy developers. Industry would have to be assured that the duration of the permit would be sufficient. This option does not directly improve the information system or integration of quantity and quality.	Selective permits applying to industry are feasible since they are less threatening to agricultural interests. Inclusive permitting faces much stronger barriers throughout the region.
Coordinated administration (Utah Plan)	This option does not exist in the West, but similar planning and administrative structures exist in Ohio and Texas.	The primary advantage is integration of water quality and quantity. In addition, flexibility can be greatly increased since different uses of water can be found in different parts of the state. Since the plan stresses information about current uses, day-to-day management can be improved. The plan also encourages public education and participation as well as political accommodation over the appropriate uses of water.	Since this option emphasizes accommodation of interests, it is likely to develop slowly. Since water is regarded as a property right in the West, the plan is essentially useful in areas where not all rights have been perfected.	Substantial opposition from irrigated agriculture can be expected. However, as demands for water continue to increase, particularly from more economic water uses, this plan could be very attractive to encourage flexibility of water use and to accommodate growth.

bigger role in how water is used. Removal of at least some of these institutional constraints appears to be a prerequisite for water rights exchange systems.

In part, choice among alternatives will be an economic decision. Since industrial users, particularly energy, are generally able to pay substantially more than agriculture for water rights and water transportation systems, pressures will increase for states to facilitate water transfer. Such transfers could be very beneficial economically to individual farmers. Water banks facilitate such transfers most directly by establishing a mechanism for coordinating transfers. The general trade-off associated with alternatives to promote economic efficiency is the probability of significant changes in the economy of the West. In this regard, costs include loss of income to the agricultural sector, reduced agricultural production, and the possibility of water rights speculation which would increase water prices.

Each alternative, except time-limited permits, directly improves the information base about water resources. The "Utah Plan," an example of coordinated administrative systems, most directly addresses information needs by integrating water quality and availability information for hydrologic units. A water exchange bank could serve as a clearinghouse for information on water quality and availability. However, both reducing institutional constraints to transfers and establishing water banks represent a longer-term choice for improved information.

A central weakness of current state administrative systems is that water availability and quality are the responsibility of different state agencies. This makes consideration of the interrelationships between the two more difficult. It also can promote direct conflict, as is typified by situations in which water quality agencies must make requests to water appropriation agencies to preserve minimum stream flow. Only one of the alternatives discussed in this chapter, coordinated administrative systems, is specifically designed to address this problem. In the "Utah Plan," for example, this would be accomplished at the local level (hydrological units) and at the state administrative level. Coordinated administrative systems could take a variety of forms, including establishment of Departments of Water Resources which would be mandated to balance availability and quality concerns. Whatever the form, a clear need exists in every state of the study area to address this problem.

Feasibility is often regarded as a weakness of almost any change in existing administrative systems. For the alternatives discussed in this chapter, feasibility appears to be a realistic concern. This is in part because some of these options would make significant changes in laws, doctrines, or state responsibilities. Most of the alternatives would probably be opposed by agricultural interests. In the case of the "Utah Plan," such opposition was largely responsible for the failure of the plan in the legislature. However, it is also true that many of these choices have already been introduced in states facing water problems. States have changed laws or administrative policies regarding beneficial use to account for instream values (Colorado, Montana, New Mexico, and Arizona) or to allow transfers (Idaho). Water banks have been adopted in Idaho and were used successfully in California in the 1976-77 drought. Time-limited permits have been established in Florida, Iowa, and Utah. Because of increasing demands on water and pressures to improve management of the resources, these approaches to water administration are likely to become more feasible within the next few years.

NOTES

[1]Reservoirs affected include Ft. Peck in Montana; Sakakawea in North Dakota; Oahe, Big Bend, and Fort Randall in South Dakota; and Saving Point in Nebraska (U.S., Congress, Senate, Com. on Int. and Ins. Affairs 1975).

[2]Colorado law places a restriction on water rights transferred from agriculture to oil shale: withdrawals can only occur during the irrigation season, which amounts to three months of the year (June-August) (Fischer 1979).

8
Regional and Federal Roles in Water Management

INTRODUCTION

As discussed in Chapter 2, jurisdictional disputes over western water management are critical issues which are aggravated by the potential for large-scale energy development. Jurisdictional disputes are also intensified by changing perspectives on the federal role in water reclamation projects and by increased demands for water from a wide variety of users. Because these changes raise fundamental questions about economic development in western states, such as agricultural versus energy priorities, the resolution of jurisdictional disputes is likely to be an evolving, ongoing process.

Both the federal government and state governments have legitimate interests in western water resource management. In fact, the history of western water resource management in this century has been one of substantial agreement between the federal government and the states on how scarce water resources should be used and about what the appropriate federal role should be. Now, considerable disagreement exists among federal and state policymakers and other stakeholders. Thus, an important question in dealing with water problems concerns the nature of the future federal role in water resource management.

The purpose of this chapter is to assess alternative roles for regional organizations and the federal government in the management of western water resources. The following section identifies major trends in state-federal relationships and develops several criteria for evaluating alternatives. This is followed by discussions of three general categories of alternatives: federal assistance to states; changes in federal policies; and cooperative state-federal management (see Table 8-1). The advantages and disadvantages of the specific alternatives identified in Table 8-1 are assessed and the final section of this chapter provides a summary and comparison of alternatives.

231

TABLE 8-1:   Alternatives for Regional and Federal
             Management

Category	Specific Alternative
Technical Assistance	Baseline Information Systems
	Coordinated Grant Support
Changes in Federal Policies	Quantification of Reserved Rights
	Federal Water Pricing
Cooperative State-Federal Management	Federal-Interstate Compact Commission

Some of the alternatives included in this discussion
may not be regarded as feasible given current circumstances. However, because circumstances are changing in
the West and because energy and natural resource policymaking is subject to diverse influences, present per-
ceptions of feasibility may not be accurate in the
future. For example, if the United States faces another
oil embargo or if foreign oil sources continue to be
subject to significant uncertainty, then pressure may
increase for a stronger federal role in western water
management in order to enhance domestic energy supply.
Thus, the discussion presented in this chapter is
intended to address a range of alternatives for federal-
state relationships in water management over the next
several decades.

EVOLVING STATE-FEDERAL RELATIONSHIPS

Water resource management in the West has always
been primarily a state responsibility and, certainly,
states do not want to lose their authority and control.
Water issues are central to questions of future economic
growth and development in the region; thus, it is under-
standable that states strongly resist attempts by the
federal government to change the nature of state-federal
relationships in this area. As a result, concerns about
water often are raised by state officials as much to resist federal control as to try to resolve specific issues
related to water availability or water quality.

On the other side of this state-federal conflict, pressures for a stronger federal role in water management continue to increase. A strong federal role has been justified by a variety of factors, including the high percentage of federal land ownership in the region, national pressures to preserve the West's scenic and recreational areas, and the fact that so many of the West's water projects are a direct result of federal financing. However, perhaps none of these will be as important as national energy goals in influencing the future federal role in western water resource management. Given the near certainty that domestic energy production will continue to be a high priority and the fact that the West contains such large quantities of resources, it is very likely that the federal government will continue its efforts to ensure that water supplies are adequate and available to develop these resources. Recent indicators of this trend include the national water policy initiative and a declaration of federal "nonreserved rights" (see Chapter 2).

However, these activities should not be interpreted to mean that a stronger federal role is necessary or inevitable. In fact, states have several options for preempting or minimizing the prospects for more centralized federal water management. Some of these options have been examined in Chapter 7 and others will be discussed below.

As we have noted, new approaches to state-federal relations will be difficult to develop. The most significant barriers are the following:

- The current policy system, which involves state, interstate, state-federal, and international agreements and institutions, is characterized by inertia and complexity. New approaches generate substantial opposition from existing institutional forces;

- States desire control over intrastate resources. Although states depend on federally-financed water projects, they resist federal "intervention." New water projects threaten to change both the nature of financing and formal responsibility for control of the water developed by the projects; and

- Economic costs of water projects, management, and water quality controls are significant. Federal assistance and coordination could benefit both states and the federal government by reducing uncertainties about water resources.

However, states typically resist federal technical assistance and coordination because of both distrust of federal motives and the implications that states are not currently managing water effectively and efficiently.

It is not likely that any single alternative will be successful in overcoming all of these barriers to developing a better relationship between federal and state governments. However, several alternatives have been suggested for dealing with western water issues which would directly influence state-federal relationships. The following sections evaluate several of these alternatives according to the following criteria:

- Improved information: Will the alternative improve the information base about current use patterns, quality, rights, etc.?

- Flexibility: Will the alternative help states to meet multiple demands for water use?

- Effective management: Does the alternative improve day-to-day management of the resource? Does it improve long-term policy planning?

- Integration: Will the alternative improve states' ability to consider and administer water quantity and water quality together rather than separately?

- Efficiency: What are the economic costs, risks, and benefits?

- Feasibility: Can the alternative be adopted, carried out, and enforced?

TECHNICAL ASSISTANCE TO STATES

As discussed in chapters 2, 3, 6, and 7 of this book, improvements in water management are needed at the state, regional, and federal levels. An important element of improved water management is better information about water resources. While much basic data exists, it is often insufficient to address specific problem areas. More importantly, even when data does exist, it is often not tied to management systems. Thus, two of the most important needs in this regard are: improving research and development (R&D) to help meet the changing goals of water use; and integrating various R&D programs directly into state management agencies. While all levels of government have a responsibility in this area, direct water management is likely to continue to be primarily a state responsibility. This section discusses two alternatives for improving state management through federal R&D assistance: improving baseline information and providing coordinated grant support to states.

Existing Programs

Research on water availability and quality is supported by a wide range of state, federal, and private sector programs. Over twenty federal agencies conduct water management and R&D activities (U.S., National Water Commission 1973; NSF, National Science Board 1978, p. 24). The general purposes of these programs include the following (U.S., National Water Commission 1973):

- Mission research to support specific responsibilities of an agency such as the Environmental Protection Agency (EPA);

- Surveys to advance understanding of hydrologic systems and processes, such as those conducted by the U.S. Geological Survey (USGS);

- Basic and applied research grant programs sponsored by the National Science Foundation; and

- Large-scale water resources development activities designed to accomplish specific objectives.

The EPA and the Department of Interior (DOI) each conduct about 30 percent of the water-related federal research. Nearly all of the EPA research has been devoted to water quality, generally focusing on technical research such as the fate of specific pollutants in fresh water. DOI has emphasized research on water augmentation and conservation (NSF, National Science Board 1978). Coordination of water research related to energy development has been done largely by the Water Resources Council.

State water resources research is frequently conducted through land grant universities in direct support of the development and evaluation of state water plans.

For example, the Water Resources Laboratory at Colorado State University assesses water development projects to determine the economic effect of alternative water development projects, the effect of water projects and resource development plans on local and regional flows, and potential changes to water quality. At the Utah Water Research Laboratory of Utah State University, watershed models have been developed to assess the combined effects of energy and industrial development on the availability and quality of water in Utah's streams. Many of these studies support or provide information to state water planning agencies which are typically a part of a state's Department of Natural Resources.

Federal R&D assistance programs to states include the Office of Water Research and Technology (OWRT) of the DOI. The objective of this R&D program is to improve the management capacity of water resources so that adequate supplies of water are available to meet the requirements of the nation's expanding population. OWRT has provided $120,000 per year to fifty-three research centers. Research organizations performing this activity in the eight-state study area are listed in Table 8-2. As indicated in the research subject areas listed in this table, research conducted by these organizations primarily addresses basic research and technology development. The research does not currently emphasize information directly useful for managing water resources through the existing allocation system, nor is the research generally coordinated for specific state water development plans. However, the mission of OWRT could be expanded to meet these management needs.

R&D Needs

This book identifies several issue areas made more difficult by inadequate information. Among the most pressing R&D needs are the following:

Water Consumption Data. Industry, government, and independent research efforts have conflicting data on water consumption by energy facilities and on alternative process designs to reduce this consumption. Improvements in these data are critical to resolve issues about how much of the limited water supply in the West will be available for population and energy growth and agricultural interests. The factors which influence data variability need to be systematically identified and defined. Although both the EPA and the Department of Energy (DOE) fund water consumption research, these efforts

TABLE 8-2: Office of Water Research and Technology Research Centers

State	Center	Research Subject Areas
Arizona	Water Resources Research Center University of Arizona Tucson, Arizona	Water policy and scarcity, strip mine hydrology, irrigation technology, aquifer hydrology, water use and salinity control models, water pollution, floodwater management, desalination
Colorado	Environmental Resources Center Colorado State University Fort Collins, Colorado	Water pollution, soil conductivity, reservoir simulation and operation, evapotranspiration, urban stormwater modelling, stream flow models, water conservation
Montana	Joint Resources Research Center Montana State University Bozeman, Montana	Hydrologic and biologic effects of coal mining, water projects cost sharing evaluation, lake biology, groundwater policy, water pollution
New Mexico	Water Resources Research Institute New Mexico State University Las Cruces, New Mexico	Water pollution, aquifer recharge, irrigation drainage, water rights policies, surface hydrology
North Dakota	Water Resources Research Institute North Dakota State University Fargo, North Dakota	Floodplain ecology, lake biology and hydrology, water pollution, flooding effects, demography, watershed models, wildlife, evaluation
South Dakota	Water Resources Research Institute South Dakota State University Brookings, South Dakota	Wetlands utilization, soil moisture prediction, rural water system economics, water pollution, irrigation management
Utah	Center for Water Resources Research Utah State University Logan, Utah	Models of grazing and infiltration, water pollution, lake ecology, irrigation methods, culvert design, water flow and dispersion models, water policy
Wyoming	Water Resources Research Institute University of Wyoming Laramie, Wyoming	Well siting, aquifer models, urban water requirements, freshwater biology, water pollution, water development policy, economic forecasts

Source: U.S., DOI, OWRT 1980.

need to place more emphasis on economics, process design, and operating conditions. Thus, R&D programs need to identify both water requirements of facilities and the range of water consumption levels under different economic and design assumptions.

Water Consumption From Alternative Agricultural Practices. Irrigation is currently the largest consumer of water in the eight-state study region; thus, a significant potential exists for reducing agricultural consumption. Studies conducted through the Department of Agriculture need to be closely linked to information about the location of energy facilities. For example, in the energy and agriculturally intensive area along the San Juan River, the potential for a reduction in agricultural water use is significant.

Holding Ponds Assessment. Holding ponds are used in most energy conversion facilities to reduce or eliminate discharge of pollutants. As energy conversion in the West expands, the use of holding ponds will increase. Research on holding ponds should include the following:[1]

- The failure of holding ponds, including data on frequencey of failures and their significance;

- Movements of pollutants through pond liners during normal operation;

- Long-term stability of holding ponds. Research conducted by the Assistant Administrator for Environment and Safety of the DOE, focusing on uranium mills, needs to be extended to include the long-term stability and fate of pollutants in holding ponds associated with other energy conversion facilities. Research needs to begin on alternative procedures for maintaining pond surface stability and prevention of leaching into groundwater; and

- Water consumption by holding ponds. Evaporation losses from existing holding pond technology are not adequately documented. Some state agencies suggest that holding ponds may increase water consumption of

energy facilities from 10 to 20 percent
due to evaporation of waters that could
be cleaned and reused.

Predevelopment Monitoring. Considerable uncertainty exists about groundwater quality. Improved monitoring and research, which could
build on current USGS programs, would provide
essential baseline data about hydrologic behavior and help to determine standards, appropriate control technologies, and liability for
changes in groundwater quality.

Institutional Controls. Existing federal controls over water and their effect on state and
local decisions need to be documented. Little
documentation exists to monitor how policies and
procedures are actually carried out at the local
level. Domestic water use, for example, has preference over industrial use, yet it is not possible to document the shut-down of energy facilities during periods of drought such as occurred
during 1976-1977. Only limited understanding of
the actual behavior of water distribution is
available to decisionmakers. Most of the studies coordinated and conducted through the USGS
and Water and Power Resources Service have fo-
cused on technical problems, design factors,
cost benefit studies and other engineering factors. Only limited information is available on
actual water consumption patterns among competing social sectors during periods of drought.

Policy Alternatives and Evaluation

Two specific forms of federal R&D assistance are
considered in order to improve state management capacities: (1) baseline information on water use, availability and quality; and (2) coordinated grant support to
assist states in developing information for administering water resources. Discussion of these options does
not imply that water-related R&D is entirely or primarily a federal responsibility. Indeed, if states are to
retain primary water management responsibility, they are
also responsible for improving R&D and for integrating
R&D into policymaking.

Table 8-3 summarizes the advantages and disadvantages of federal R&D assistance. Several general advantages are identified. First, the cost of R&D has been a
barrier to state programs; thus, federal assistance and

TABLE 8-3 : Summary of Federal Technical Assistance

Criteria	Findings
Improved information	Substantial improvements can be made in generating knowledge for specific problem areas. Primary needs include water consumption data for energy facilities and agricultural practices, assessments of holding pond performances, predevelopment monitoring of groundwater, and the effects of institutional controls.
Flexibility to meet multiple uses	Flexibility could be improved by removing many of the uncertainties associated with water resources. However, better knowledge may also be used to restrict the use of water for some purposes which are viewed to be too costly or risky.
Effective management	Better knowledge is a prerequisite for dealing with many of the problems identified in this report. However, the success of R&D programs depends on states' willingness to integrate them directly into water management agencies.
Integration of water quality and quantity	Since many of the information needs identified in this report address the environmental consequences and constraints of various water uses, substantial improvements could be made in this regard. However, the effectiveness will ultimately depend on whether state water agencies are integrated to consider water quality and quantity together.
Efficient water use	The economic costs of this alternative do not appear substantial if incorporated into existing federal programs. However, these expenditures are risky since they depend on state institutional changes to better integrate R&D into management.
Feasibility	Federal technical assistance is more feasible than most forms of federal activity in western water management because it provides assistance rather than intervention; it can be accomplished within existing administrative frameworks; and it helps to reduce the economic costs of state management improvements.

incentives can be used to reduce this barrier. Second, several federal programs already exist; thus, an improved R&D capacity could be obtained within existing administrative structures. Third, these are "assistance" options rather than intervention options; thus, they are more feasible than many other federal alternatives.

As we have discussed, an integrated water resource information system could reduce much uncertainty about water availability, the status of existing rights, and the effects of various water uses on water quality. Thus, such an information system could contribute to other alternatives for improving management, such as the water rights exchange system described in Chapter 7. Day-to-day management of water resources would also be improved with a central data bank. A system of current accounts would reduce the burden on state water agencies. Once implemented, the system should provide an efficient basis for routine activities including applications and permit review.

Better information does not necessarily mean that states will encourage multiple uses of water or that they will use the information to address specific problems. Indeed, the success of these alternatives depends on the willingness of states to integrate information directly into the activities of water resources boards, health departments, or other agencies dealing with water resources. Some states have done this; for example, the Illinois State Water Survey works closely with the Water Resources Center at the University of Illinois. In addition to relating R&D directly to management, the effectiveness of federal technical assistance would be greatly enhanced if western states developed a more integrated management approach by administering both water availability and water quality in the same agency.

Federal grant support to state agencies also responds to the general criticism that research conducted through federal laboratories has been deficient in strengthening state and local government (CED 1976, p. 25). A "pass-through" approach of giving grants to water resource divisions in departments of natural resources would help to avoid autonomous research units in the states and make research more effective in responding to state management needs (U.S., OMB 1975, p. 10).

Both the grant assistance and data base improvement alternatives can contribute to integrated water quantity and quality management. The feasibility of a publicly accessible water-use file may be limited by concerns over the proprietary nature of water use and recognition that water rights are vested property. However, a precedent exists in property files which are kept at the county level for real estate and for mineral patents.[2] The water resources division of the USGS currently provides background data for a more responsive information system. Nevertheless, a common accounting system among the states could be resisted by current water users, such as agriculture groups, that might believe that a common information base would facilitate competing water interests.

CHANGING FEDERAL POLICIES

Two of the most important federal activities influencing western water resources concern federal reserved rights and pricing policies for reclamation projects. As discussed in Chapter 2, these activities contribute to problems of water management by increasing conflicts between federal and state governments and by increasing the uncertainty over how much water will be available to states in the future. This section discusses two alternatives for reducing these problems: quantifying reserved rights, and modifying water pricing policies.

Quantification of Reserved Rights

The reserved rights doctrine provides the federal government and Indian tribes with a claim to considerable amounts of western water. However, these claims remain largely unquantified, resulting in uncertainty about water available to states and intergovernmental conflicts among Indians, states, and the federal government. For these reasons, suggestions to systematically quantify reserved rights have frequently arisen. Currently, reserved rights are being quantified on an ad hoc basis; many federal agencies and Indian tribes attempt to quantify only when it is required as part of a stream adjudication or when their interests may be threatened by the claims of other water users.

This section evaluates administrative adjudication, as opposed to the current practice of judicial adjudication, to quantify reserved rights. In administrative adjudication, the directors of each federal agency would prepare an inventory of all reserved and appropriative water rights under their jurisdiction. Inventories would include the following information: the legal nature of such rights including priority date; water source and point of diversion (if any); the purposes, place, and quantity of use; a listing of any court adjudication of the right; and identification of the agency under which the right is administered. For Indian tribes, the Bureau of Indian Affairs (BIA) could serve as a coordinating agency, but each tribe would be responsible for identifying their rights (Kiechel 1975). An inventory office would then be established by the secretary of interior to compile records of all water rights owned or claimed by the United States. The completed inventory would be submitted to state water management officials and revised periodically. Indian tribes would submit their claims to the states. Disputes between the claimants and the states or private users could be resolved by negotiation, or as a last resort, by the courts.

Several attempts have been made in the past to quantify reserved rights. The most recent attempt was included in the Carter Administration's water policy proposals. The BIA, for example, was directed to formulate a plan to quantify and inventory Indian reserved rights (Yale Law J. 1979). Other federal agencies were directed to quantify federal reserved rights based on actual federal needs, rather than on the full possible extension of all possible claims (Simms 1980). The entire process was expected to take up to 15 years. These quantification proposals have not been finalized.

Previous efforts to quantify reserved rights have failed in part because enough water has generally existed to meet various demands. However, they have also failed because of the difficulty in determining the quantity of water appropriate for fulfilling the purposes of the reservation and the appropriate institutional mechanism for determining these quantities. Three criteria representing different degrees of quantification have been suggested:

- Open-ended settlement, granting Indians or federal reservations current claims and providing for future additional claims, as needed;

- Final determination, treating any current Indian/federal claim liberally with the understanding that future additional demands would be relinquished; and

- Irrigable acreage, granting Indians water sufficient only to irrigate the cultivatable acreage on the reservation.

Evaluation of Quantification[3]

The benefit of quantification is that it would reduce one of the major uncertainties affecting western water availability. In contrast to the current system, rights could be determined in a relatively comprehensive manner. Once accomplished, quantification would be beneficial to a variety of potential water users, at least in reducing some of the risk currently associated with water rights.

As summarized in Table 8-4 quantification also has several disadvantages. Among the most significant of these will be the political conflicts associated with choosing criteria for quantification. Criteria will vary for each type of reservation, and little agreement exists between state and federal officials as to the applicable criterion for each type of reservation (Valantine 1980). An irrigable acreage criterion would be to the states' advantage, but Indians would almost certainly challenge this as contrary to Supreme Court rulings. Such challenges would weaken the effectiveness of quantification in reducing uncertainty. In addition, it is unclear how an irrigable acreage standard would affect Indian water rights if Indians brought greater amounts of land into production after water rights had been adjudicated (U.S., GAO, Comp. Gen. 1979, pp. 86-89). If an open-ended criterion were adopted, substantial state objection could be expected since liberal interpretation of Indian claims could mean that water available for state appropriation would be reduced more than with other criteria. Further, since open-ended settlements allow for future additional claims, this criterion still leaves uncertainty about reserved rights. Thus, of the three criteria identified above, "final-determination" appears to be the most effective choice in reducing uncertainty.

Administrative adjudication would also encourage federal officials to maximize their claims to reserved rights and place the burden of disproving a claimed right on state officials and private persons. Since the process would be a federal administrative procedure, the federal courts would hear appeals from their decisions rather than the usual procedure of state court determination of water rights. It is also not clear whether administrative adjudication would reduce litigation, since the administrative judges' decision may be contested in court (Valantine 1980).

One advantage of quantification is that it could require considerably less time and money than the present system. The current ad hoc method for federal reserved rights is expected to take at least thirty-five years and $225 million. On the other hand, quantification of all federal water rights in a comprehensive manner could be accomplished in fifteen years at a cost of $120 million (U.S., President's Water Policy Implementation 1979).

Although not legally required, compensation for users who cede a water right or suffer a hardship when other parties are granted their rights may be desirable. Compensation can work both ways. For instance, many non-Indians are currently using water to which tribes may in the future assert a right. Although the Indians

TABLE 8-4: Summary of Quantification of Reserved Rights

Criteria	Findings
Improved information	Administrative adjudication could result in a significantly improved information base valuable to the federal government, states, Indians, energy developers, and other stakeholders. However, final determination of reserved rights and compensation could take fifteen years.
Flexibility to meet multiple uses	Flexibility may be enhanced if the adjudication is successful in reducing existing uncertainties. However, some criteria for quantification may make large grants to reserved right holders which would reduce state flexibility in meeting multiple needs.
Effective management	Effective management of water resources would be enhanced by reducing risks and providing an up-to-date inventory system.
Integration of quality and quantity	No direct effect.
Efficient water use	This alternative does not promote one water use over another. However, if successful, quantification could facilitate state policies to promote efficient water use.
Feasibility	Feasibility depends on the criterion chosen for quantification. Although any quantification system would be costly, acceptance by stakeholders will depend on how well various interests are balanced. Among the criteria considered here, "final-determination" appears to be most likely to succeed.

are acknowledged to be senior right holders, the fulfill-
ment of that right might work a hardship on persons currently using the water. Thus, the government could
choose either to find an alternative source of water for
current users or compensate them for their economic loss
(U.S., National Water Commission 1973, p. 482). Such
compensation costs could be considerable. For example,
the cost of buying out farming operations in the Welton-
Mohawk irrigation district of Arizona is estimated at
$332 million, just to provide water to five central Ari-
zona tribes (DeConcini 1977, p. 41). On the other hand,
some tribes hold water rights which they are not currently exercising, and they may have no need for that
water in the near future. In such cases, the tribes
might lease their right to other users for a limited
time. It is not clear, however, that transfers to off-
reservation users would be legal (Valantine 1980).
    As this discussion suggests, quantifying reserved
rights will be difficult because of the intergovernmental conflicts involved and because of the potential

effects on current right holders. Indian tribes appear
to be in a strong legal position as senior right holders
in many areas of the West; thus, if Indian rights are
interpreted liberally by administrative adjudication, the
net result may be increased rather than decreased political conflict. If administrative quantification is not
accomplished in a manner satisfactory to Indians, then
the most likely possibility is for continued piecemeal
judicial adjudication of reserved rights.

Federal Water Pricing

The Reclamation Act of 1902 gave BuRec a general mandate to provide water storage projects to irrigate arid
lands and encourage economic development in the West.
The act required repayment of reimbursable cost by the
beneficiaries. However, because farmers could not afford payments during the depression, the Reclamation
Project Act of 1939 varied the price of water according
to the irrigator's ability to pay. An irrigator's ability to pay is based on factors such as land productivity,
types of crops, and farm size. After estimating farm
income, the ability to pay for water is figured by de-
ducting a family allowance and the cost of farming inputs other than water. In order to encourage irrigation,
the price of water has historically been about 75 percent of the ability to pay (Davis and Hanke 1971;
Kauffman 1977).
Federal water pricing and project evaluation policies
have been criticized for discouraging efficient water use
by providing artificially low water prices and discount
rates to agriculture (Howe 1977; Pring and Tomb 1979;
Hanke and Anwyll 1980). An alternative to the present
system is to base federal water prices on the real cost
of making water available rather than the ability of the
user to pay. This "real-cost" policy would include a
discount rate to reflect the opportunity cost of public
funds. Opportunity costs are defined as the return on
private investments foregone by making public expenditures. In this alternative, the price of making water
available for use is discussed, but not the price of the
water right.
The method to determine costs of storage projects is
very important to "real-cost" water pricing and the discount rate is of particular importance. The rate used
recently (6-7/8 percent) has been criticized for being
too low and thereby underestimating the true costs of
the project. Hanke and Anwyll (1980) suggest that a discount rate of 8-1/2 to 10-1/2 percent would more accurately reflect the real opportunity cost of public funds.

Evaluation of "Real-Cost" Pricing

Table 8-5 summarizes the advantages and disadvantages of "real-cost" pricing. The value of real-cost pricing depends on how much economic efficiency is valued compared to other criteria. In essence, the value of this alternative depends on whether and how much western irrigated agriculture should be subsidized. Traditionally, "ability-to-pay" pricing has been one mechanism for carrying out an explicit federal policy of supporting agricultural development. Thus, pressures to change "ability-to-pay" pricing largely reflect changing values about western development and water use.

Although estimates of the magnitude of the subsidy to irrigated agriculture differ considerably, it is apparent that water prices do not reflect the cost of making water available. One study estimated the cost at $42 per acre-foot per year (AFY) for making water available at the dam of the proposed Narrows Unit in Colorado (Howe 1977), but WPRS has recommended that irrigators be charged only $6 per AFY for this water. Thus, the subsidy for water at the dam would be about $36 per AFY and at the headgate about $69 per AFY, since about half of the water is lost between the dam and the farmers' headgates (Howe 1977). Figuring project costs with discount rates that reflect the real opportunity cost of public funds would also substantially alter the results of benefit-cost studies of many proposed projects (Hanke and Anwyll 1980).

One problem with "real-cost" pricing of federal projects is that other values of projects are often ignored. For example, it has been estimated that an "average of 84.4 percent of federal capital investment is reimbursable to the federal treasury by beneficiaries in areas of projects: water users, power users, etc." (WESTPO 1978, p. 35). It has also been estimated that $11 billion in increased business activity resulted from all reclamation programs throughout the nation. This increase added $1.8 billion in tax revenue to the federal treasury, and $900 million to state and local governments' revenue. Western Governors' Policy Office estimated that flood control benefits between 1950 and 1976 averaged $48.1 million annually (WESTPO 1978, p. 35). Accordingly, the Council of State Governments has voiced support of the current method of figuring discount rates that appears to be very beneficial to the project states (Wilson 1978).

"Real-cost" pricing has been strongly supported by environmental interests, such as the Environmental Defense Fund (EDF). Environmental support is typically based on opposition to additional water projects in the West. It has been argued that such projects would:

TABLE 8-5:   Summary of "Real-Cost" Water Pricing

Criteria	Findings
Improved information	No direct effect.
Flexibility to meet multiple uses	Flexibility would be increased if project water were made available to the highest bidder. Energy interests would be greatly favored over agricultural interests.
Effective management	This alternative would not necessarily improve management.
Integration of quality and quantity	Relationship between quality and quantity is not necessarily addressed. Quality of water may be improved over long term as agricultural production decreases.
Efficient water use	Primary value of this alternative is in promoting economically efficient water uses. Industry and municipal interests would be able to afford "real-cost" water pricing, but many agricultural buyers would not. However, economic efficiency calculations ignore many indirect benefits of water projects.
Feasibility	Substantial opposition exists to changes in pricing policies, particularly from agricultural interests and states which stand to benefit from future projects.

    (1)   Contribute to the loss of wildlife habitat;

    (2)   Have adverse effects on salinity concentration in watercourses; and

    (3)   Have uncertain cumulative effects on salinity and wildlife, since comprehensive environmental impact statements (EIS) for each affected river basin have not been prepared. WPRS recognizes the need for a comprehensive EIS to include the cumulative impacts of individual water projects on the environment, but has indicated that it will not halt construction in progress until such an EIS is completed (El-Ashry and Weaver 1977).

The EDF also contends that water conservation is the alternative to building new dams and reservoirs, suggesting that water is largely wasted through the ability-to-pay pricing criterion.

Recipients of the agriculture subsidy oppose efforts to change pricing policies. Governor Lamm of Colorado has stated that proposals to change "ability-to-pay" pricing would encourage industrial water development, thus further transforming the West into an "energy colony" for the industrialized East and Midwest (<u>Denver Post</u> 1979). Harris Sherman, former director of the Colorado Department of Natural Resources, has said such proposals would effectively halt further construction of WPRS projects, and "reward those interests that can pay the highest price for water." Further, Sherman contended the plan "would destroy those uses most compatible with the environment" (<u>Denver Post</u> 1979).

In addition, agricultural interests strongly object to allegations that agricultural subsidies in the form of ability-to-pay pricing promote wasteful water use. In spite of the incentives, they argue that most farmers realize supplies are finite and use water judiciously. They also argue that most of the low-cost water is in use, so that most farmers who do not receive water from a federal project must supplement it with other more costly sources, such as groundwater. Thus, any subsidized water that is wasted must be replaced by more costly water. Even though federal water is subsidized, it still represents a major expenditure for farmers, who are not likely to want to pay for more than they need.

These conflicts among energy, environmental, and agricultural interests reduce the feasibility of changing federal water pricing policies. More importantly, "real-cost" pricing primarily affects future rather than present users of federal project water and would not influence the high percentage of agricultural users relying on privately developed water. Thus, this alternative is not a comprehensive choice for changing water management. It is apparent that "real-cost" pricing could substantially reduce future agricultural use of water. To what extent this would make more water available for energy, municipal, or other uses would depend on many other elements of the water policy system, including questions over the control of water from federal projects and the priorities of state water policy, including the transferability of water rights (see Chapter 7).

JOINT STATE-FEDERAL MANAGEMENT

Because many western water problems are regional or national in scope, a variety of mechanisms has been suggested to provide a cooperative approach to decision-making. These mechanisms include river basin commissions, federal water corporations, and federal interstate compact commissions. Emphasis is placed on the compact commission as a potential mechanism to improve river basin management.

The Missouri River Basin Commission (MRBC) was established by the president according to Title II of the Water Resources Planning Act of 1965. Title II commissions have no management authority; their primary function is to coordinate planning efforts between the states and the federal government (Derthick 1974). Commission members include representatives from states, the various federal agencies involved in water resources activities, and from interstate compact commissions (Yellowstone and Big Blue River Commissions). A chairperson responsible for coordinating federal viewpoints is appointed by the president.

The federal corporation is frequently used to perform complex financial or operational tasks for the federal government. Such corporations are chartered by the federal government and retain some degree of continuing federal control. In water management, only one prototype exists for a federal corporation--the Tennessee Valley Authority (TVA). Efforts to duplicate the TVA in the Missouri, Columbia, and other basins have failed. The federal water corporation requires both federal legislation (to authorize federal representatives) and implementing legislation by each of the participating states. Because of the difficulty in accomplishing this, particularly in the West, and because of other political barriers to forming such a cooperative mechanism the National Water Commission concluded that the federal water corporation is unsuited to either regulatory or comprehensive river basin responsibilities (1973, p. 431). The Commission recommended that future federal-state interstate compacts should be drafted to allow chartered corporations to perform "discrete operational tasks," such as operation or maintenance of particular projects (1973, p. 431).

The remainder of this section discusses and evaluates the federal interstate compact commission as an alternative form of joint federal-state management. This option appears to be both more appropriate to the situation in the western states and more feasible than expanding the authority of the river basin commissions or creating a federal water corporation.

Federal-Interstate Compact Commission

The powers of a federal-interstate commission would be much more comprehensive than those of a Title II commission. For example, the Delaware River Basin Commission (DRBC), the first federal-interstate compact commission, which was established by the Delaware River Basin Compact (1961), is required to:

* Develop and effectuate plans, policies, and projects relating to the water resources of the basin;

* Adopt and promote uniform and coordinated policies for water conservation, control, use, and management in the basin; and

* Encourage planning, development, and financing of water resources projects according to such plans and policies (Derthick 1974).

In addition, the DRBC has the power to construct or own any projects or facilities necessary to fulfill the function of the compact. However, it does not have the power to abolish existing programs or take over the management of existing projects.

The membership of the DRBC consists of the governors of each of the signatory states or their designates, and a federal member appointed by the president. Each member has one vote, a majority of members being required for decisions. Besides allocating the waters of the Delaware River among New York, New Jersey, Pennsylvania, and Delaware, a major function of the DRBC includes formulating a comprehensive plan for water resource development and management in the four states, binding on all five member governments (Derthick 1974). The president, however, may suspend, modify, or delete any provision of the comprehensive plan as he deems necessary to the national interest. Concurrence of the federal member with the comprehensive plan is assumed unless a notice of nonconcurrence is filed within sixty days of the action (Delaware River Basin Compact 1961). No public or private water facilities may be constructed unless consistent with the plan.

A third important function of the DRBC is in controlling pollution. Article V of the Compact gives the DRBC the authority to take measures to control existing and potential pollution by constructing projects, making surveys, and establishing rules and regulations for sewage and industrial waste disposal. The federal-interstate compact commission form of regional water management has also been established in the Susquehanna River Basin (Susquehanna River Basin Compact 1970), and has been proposed for the Potomac River, the Great Lakes Basin, the Hudson River, and the Missouri River Basin (Muys 1971).

Evaluation of Federal-Interstate Compact Commission

Before inception of the Compact, the Delaware River Basin experienced many conflicts over water resources, especially in water allocation. For example, conflict between Philadelphia and New York City continued from the early 1920's onward regarding Delaware River allocations. It was a period marked by repeated litigation among New York, Pennsylvania, New Jersey, and Delaware. Similar conflict existed over the allocation of pollution control burdens along the river. It wasn't until the Compact was in place that improvements in water apportionment, management, and quality were implemented. Since then, the DRBC has been able to improve the policymaking environment. During its early years it successfully mitigated the effects of the 1965-67 drought in the Northeast and extended its jurisdiction to include control over groundwater in the basin. It has planned for the management of several water augmentation projects, established basin-wide water quality regulations and abatement schedules, and participated in the regulation of power plant siting and industrial development. In general, water quality has improved and the demands of numerous large-scale water users have been assimilated. In fact, the Commission proved so effective at coordinating water management and policymaking that the water agencies of the four state members were reorganized to complement Commission operations (U.S., National Water Commission 1973, p. 423).

Table 8-6 summarizes the advantages and disadvantages of the federal-interstate compact for the West. Probably the most significant advantage is the coordinating and comprehensive planning authority of the compact commission. Unlike the planning activities of Title II commissions, such as the MRBC, the comprehensive plan of a federal-interstate compact commission would be binding, subject to a veto of the federal member. Title II commissions are also authorized to formulate cooperative basin-wide plans, but they are nonbinding.

The importance of a federal-state partnership in water resources management has been emphasized by four national commissions (U.S., President's Water Resources Policy Comm. 1950; U.S, Comm. on Organization of Executive Branch of Govt. 1955; U.S., Comm. on Intergovernmental Relations 1955; U.S., Congress, Senate 1961; cf., U.S., Congress, Senate, Com. on Judiciary 1961). These commissions found that necessary conditions for successful water resource management were: (1) a unified federal viewpoint; and (2) cooperation among state and federal agencies. The historically large role of the federal government in water resource management in the

TABLE 8-6:   Summary of the Federal-Interstate Compact

Criteria	Findings
Improved information	Information is enhanced by the coordination and comprehensive planning function of the compact commission.
Flexibility to meet multiple uses	Flexibility to meet regional or basin-wide needs is the primary advantage of this mechanism. However, this advantage is dependent on the willingness of states to agree to delegate authority to the compact commission.
Effective management	Overall management capacity can be considerably improved only if the commission is given broad authority. If this occurs, individual states' authority is lessened, and federal influence is superceded by regional influence. State water policies must comply with the plan of the compact commission, which has binding authority.
Integration of quality and quantity	Integrated planning is a primary mission of the commission, particularly regarding federal agencies. Intergovernmental coordination can be enhanced.
Efficient water use	This alternative is not directly responsive to promoting efficient use of water.
Feasibility	Under current conditions, interstate and state-federal conflicts, and conflicts within federal water agencies substantially reduce the feasibility of this choice. If these conflicts are reduced in the future, this option appears to be the most feasible cooperative formal mechanism for joint federal-state water management. States retain considerable day-to-day authority and also have a strong voice in the commission.

eight-state area makes integrated planning particularly appealing.

The National Water Commission (1973, p. 424) advocated the federal-interstate compact as the preferred mechanism for water resource planning and management in multistate regions. The Commission concluded that the chief advantage of this approach is its adaptability to the particular needs of a basin:

...it can be shaped to meet any problems the States desire, in accordance with the particular regional philosophy of appropriate intergovernmental relations. Thus, it can be targeted on a single problem, such as water quality management, or may seek comprehensive, multipurpose goals [1973, p. 424].

Under present circumstances, the appropriateness of this idea to the West is questionable, since the philosophy of intergovernmental relations in the area includes a strong component of independence. Thus, the political difficulty of generating agreement among several states, particularly the seven states of the Colorado River Basin, appears to be the most significant barrier to this mechanism. The National Water Commission referred to this barrier as the exceedingly long time required to negotiate such agreements. The federal-interstate compact, as experienced in the DRBC, has also been criticized for not fully using its broad water development authority. Although the DRBC has provided an important coordinating function, it has not been as successful in directing future growth in the basin and has been overly concerned with immediate problems at the expense of long-term management (Derthick 1974; Muys 1971).

The authority of a federal-interstate compact commission to coordinate state and federal water development efforts and to integrate water quality and water quantity problems would greatly improve the efficiency of policymaking and implementation. Day-to-day management would be left to state and local organizations; nonetheless, their water agencies would be bound to the provisions of the comprehensive plan. This alternative would be likely to reduce the flexibility of separate executive branch agencies, since more consultation and coordination would be required. The flexibility of the state water management authorities would be increased, however, due to the enhanced potential for influencing the policies of the federal executive branch.

SUMMARY AND COMPARISON

The water resource situation in the West is changing: increased demands from a variety of users, increased concern about water quality, and jurisdictional disputes are among the most important pressures on the existing system. As a result, many suggestions have been made for improving water resource management on a regional or federal level. This chapter has discussed several of these alternatives: federal technical assistance to states, quantification of reserved rights, "real-cost" water pricing, and federal-interstate compacts. A summary of findings is presented in Table 8-7. The strengths and weaknesses of these alternatives depend on the criteria chosen to evaluate them; the six criteria used in this chapter represent six of the questions which we have found to be important to water resource management.

Among the most important factors affecting the out-
come of this evaluation is the uniqueness of the water
situation in the eight-state study area compared to
other regions of the country. This uniqueness includes:
(1) the richness of energy resources which are largely
located in water-scarce areas; (2) the high percentage
of federal lands throughout an area which is well known
for the strength of its feelings of independence from
"outside intervention"; and (3) the complexity of the
water policy system which includes international treaties, a variety of interstate agreements, diverse federal regulations and programs, and the intricacies of
state appropriation systems. Because of these and other
factors, two general conclusions can be drawn. First, no
simple solutions to future water management problems
exist--a mix of federal, state, interstate, and public
sector initiatives will be required. Second, management
approaches tried in other states, regions, or countries
are unlikely to succeed in the West without significant
revisions to fit the circumstance of the West.

Of the alternatives discussed in this chapter, the
ones which appear the most promising might be termed
"cooperative" approaches to water resource management;
that is, they recognize the legitimate state and federal
interests in the region. In this regard, federal technical assistance in developing water information systems
and federal-interstate compacts have several advantages.
Among the most important of these is their potential to
improve the information about water availability, existing uses, and relationships between quantity and quality. As discussed throughout this chapter, improving
the information base in these areas is a prerequisite
for several other management goals. The advantage of
federal technical assistance is that it can increase
states' day-to-day management capacity with a minimum of
federal "interference." The federal-interstate compact
provides a more integrated approach to developing water
resource plans, in addition to giving the states a
strong voice in the compact commission and virtual control of day-to-day management.

Technical assistance also rates more favorably on
the feasibility criterion than do the other alternatives
discussed in this chapter. This is because assistance
does not threaten the states and could be implemented
through existing federal programs. The federal interstate compact approach raises more questions. The most
resistance to this idea may come from the variety of federal agencies with responsibilities in the study area,
as their activities could be substantially affected by a
federal-interstate compact.

Quantification of federal reserved rights and "real-
cost" pricing for water from federal projects received
considerable attention during the Carter Administration.

TABLE 8-7: Comparison of Regional and Federal Management Alternatives

Alternative	Current Status	Advantages	Disadvantages	Feasibility
Federal technical assistance	Over 20 federal agencies conduct water management and R&D activities. R&D assistance to the states is primarily channeled through the Office of Water Research and Technology.	Primarily, this option is beneficial in improving the information base about water situations and problems. Better knowledge is a prerequisite for dealing with many of the issues identified in this book. The cost of this option is relatively low.	The success of this option depends on the willingness of states to integrate water quality and availability concerns and to directly link R&D programs to policy agencies.	Federal technical assistance is more feasible than most forms of federal activity since it provides assistance with minimal intervention. It can also be accomplished within existing administrative frameworks.
Quantification of reserved rights	Several attempts have been made to quantify reserved rights administratively. Most recently, the Carter Administration included adjudication as part of its water policy proposals. These proposals were not finalized.	A significantly improved information base could result, valuable to the federal government, states, Indians, energy developers, and other stakeholders. Risks of development could be reduced if a final determination were made. Depending on the criterion state flexibility could be enhanced.	No acceptable criterion has been found to serve as the basis for quantification. If reserved rights are interpreted liberally, the net result could be increased political conflict. If interpreted strictly, continued judicial interpretation would be sought by Indians. This is of little short-term benefit.	Acceptance depends on how well competing interests are balanced. Of the criteria being considered, "final determination" appears to have the best chance of succeeding.

TABLE 8-7: (Continued)

"Real-Cost" water pricing	Several proposals have been made to base the price of federal water on the real cost of making it available, rather than the ability of the user to pay. Not one of these proposals has been successful.	This proposal would help to make future (unallocated) water available for nonagricultural interests; primarily industrial; thus, it would be advantageous to the energy industry. Economically efficient water uses would be promoted.	This option does not affect present water users; thus, it is a limited choice for dealing with many water related problems. It does not directly enhance state management and it may ignore many indirect benefits of water projects.	Substantial opposition exists, particularly from agricultural interests and states which would benefit from future projects.
Federal Interstate Compact Commission	Although recommended by several presidential commissions and the National Water Commission, only two prototypes exist-- The Delaware River Basin Commission and the Susquehanna River Basin Commission.	Overall management capacity could be significantly improved if the commission were given broad authority. If this occurred, individual authority would be lessened but regional authority would be increased. This option would greatly facilitate integrated management and gradually improve the information base about water problems.	The primary disadvantage is that this option requires legislative consent of each state involved and each state must be willing to grant the commission authority to deal with problems of the basin.	This is generally considered to be infeasible under present conditions, but may be increasingly feasible unless states adopt other management strategies to improve water resource policymaking.

However, both alternatives face strong barriers. For quantification, there is no evidence that an acceptable criterion for quantification can be found which will satisfy states, Indians, and the federal government. If an acceptable criterion is found, administrative adjudication is a long-term (15 year) process. "Real-cost" pricing for reclamation water addresses a very limited goal of forcing future water users to pay the cost of making water available. If achieved, this will not have a widespread influence in the West since so much water is already being used, primarily by the private sector. While frequently justified as discouraging wasteful water use, its more likely effect will be to promote nonagricultural uses of future federal project water. If this occurs, energy developers are the likely beneficiaries.

NOTES

[1]Some of these questions are being addressed by EPA-funded research, particularly the National Surface Impoundment Assessment Program.

[2]The location-patent system allows a company or individual who discovers a mineral deposit (iron, copper, uranium, or other "hard rock" minerals) of profitable quantity to file a mining claim which gives the prospector exclusive title to both the mineral and surface rights. This is called a mineral patent. The patent does not require that the deposit be mined, or that the developer pay for any minerals extracted and land used, or that mined land be reclaimed (Choitz 1978, p. 27).

[3]"Quantification" refers to administrative adjudication in the remainder of this section.

9
Summary, Conclusions, and Comparison with Other Studies

PURPOSE AND SCOPE

This book assesses the relationship between water resources and energy development in eight western states. It has two general purposes:

* To identify the most important water impacts and issues likely to be associated with energy development; and

* To identify and evaluate policy alternatives for dealing with problems and issues.

Although the study was sponsored by the Environmental Protection Agency (EPA), it is expected that the findings will be useful to a wide range of interests, including other federal agencies, the Congress, state and local governments, environmental interest groups, energy companies, and Indians.

Two additional characteristics of the study are important to the findings presented below. First, the report is future oriented but not predictive. That is, its purpose is not to predict exact levels of western energy development, water use demands, or other conditions over a specified time frame, but rather to look at a range of questions about water and energy development likely to be important over the next several decades. Thus, much of the analysis is structured to anticipate possible issues by asking "if-then" questions about energy development and water resources.

Second, this report takes a broad and integrative approach to the relationship between energy and water resources. This requires that environmental, social, economic, legal, and technical perspectives be considered. Our study report can be contrasted with some other publications on this topic which are limited to the question of whether enough water physically exists to support western energy development. This question is only one of several important questions which need to be considered together. Indeed, attention only to the physical supply of water without attention to the significant institutional constraints and political issues involved can easily lead to an oversimplified view of water and western energy.

The following two sections briefly summarize our findings. A final section then compares the findings of this report to those of two other recently published reports on water and western energy.

IMPACTS AND ISSUES

Energy development will create several impacts on western water resources. The nature and severity of the impacts will depend on the technologies chosen to develop resources and the locations where they are sited. Among the most important factors are:

• The type and number of facilities;

• The cooling technologies;

• Effluents produced by the facilities;

• Water availability region-wide and at specific development locations; and

• The location and depth of aquifers in relation to mining.

Many of the conflicts over water availability and quality discussed in this report will exist even without energy development because the region is generally water-short and because a variety of demands are being made for the use and protection of the resource. Not only will energy development aggravate these issues but they, in turn, can directly influence energy resource development decisions. Eight issues associated with energy development are emphasized in this report and summarized below.

Consumptive Water Requirements for Energy Development

Table 9-1 indicates water consumption for the energy facilities considered in this study. In estimating future consumptive water requirements for energy development in the region, considerable uncertainty exists in several areas:

TABLE 9-1: Consumptive Water Requirements for
            Energy Facilities

Facility and "Standard" Size	Water Consumption per Facility (AFY)
Coal-Fired Power Plant (3,000 MWe)	23,900-29,800
Coal Gasification (250 MMscfd)	4,890-8,670
Coal Liquefaction (100,000 bbl/day)	9,230-11,750
TOSCO II Oil Shale (100,000 bbl/day)	12,900-18,600
Modified In Situ Oil Shale (100,000 bbl/day)	7,600
Uranium Mine and Mills (1,000 mtpy)	270-300
Slurry Pipelines (25 MMtpy)	13,500-18,400
Surface Coal Mines (25 MMtpy)	Neg.-1,240
Geothermal (100 MWe)	12,700-13,700

AFY = acre-feet per year   mtpy = metric tons per year
MWe = megawatt-electric   MMtpy = million tons per year
MMscfd = million standard cubic   Neg. = negligible
         feet per day
bbl/day = barrel(s) per day

- There is a lack of experience with commercial-
  size facilities for new technologies such as
  oil shale and coal synfuel processes;

- The type of cooling process chosen can create
  significant differences in water consumption;
  and

- There are inherent difficulties in estimating
  future energy development levels in the region
  over the next twenty years and beyond.

Because of these and other questions, precise estimates
of the extent to which water availability will affect
energy development are not possible. However, it is
clear that enough water physically exists on a region-
wide basis to support very large levels of energy development. For example, the total consumptive water requirements for our Nominal Demand scenario, representing
a very high level of development by the year 2000, would
require a maximum of 22 percent of the remaining unused
water supplies in the Upper Colorado and Upper Missouri
river basins. Nevertheless, water availability and
quality issues are certain to affect the overall level,

location, and types of energy development that will
occur. Even if water shortages are not a concern region-
wide, water is relatively scarce in many energy-rich
regions, including the Powder River Basin in Wyoming and
Montana, the oil shale area of western Colorado and
eastern Utah, and the Four Corners area of New Mexico.
In addition, a variety of institutional and political
questions will determine the amount of water made available to supply new energy projects. While physical
availability will be an important factor, it may be less
important than competing demands from other users, policies protecting environmental quality, and increasing
conflict among federal and state governments over water
resource management. The importance of these kinds of
constraints is indicated by the difficulties slurry
pipeline projects have encountered in obtaining needed
water.

Increasing Demands for Water Use

Although agriculture is the dominant water user in
the study area, there is a rapidly growing demand for
water from other users, including energy producers,
municipalities, Indian tribes, and environmental interests. Each of these demands can influence the level of
agricultural production in the region and, thus, threaten
many existing interests and values which have developed
within a predominantly agricultural economy. In addition, disputes among energy, environmental, Indian,
agricultural, municipal, and other interests set an increasingly complex context for water policymaking. In
some areas, demands made by these various interests mean
that not all potential users can be satisfied. And in
other areas, competing demands add to the uncertainty
and delay which can restrict the level and location of
energy development.

Uncertainty and Complexity of the Water Policy System

Managing water resources is made more difficult by
the complex combination of state water law, federal water policies, court cases, interstate agreements, and
international treaties which determine how water will be
used. Although this system has dealt successfully with
many difficult water problems, it also creates barriers
to change, discourages a diversity of water uses, and
provides few incentives for efficient management and
conservation of the resource. One of the most serious
deficiencies of this system is that a state appropriation system typically settles disputes over water use
only after damage of a water right has occurred. Thus,

the current system is designed primarily to react to
questions about the legality of rights and uses and not
to manage the resource on a day-to-day basis. A second
major deficiency is that various components of water
resource management are typically treated as if they
were unrelated. The most obvious example is that the
water allocation system in most western states is not
tied directly to water quality protection. A third
deficiency is that change is largely discouraged. This
can be attributed to the long and complex history of the
water policy system and to the reliance on court rulings
to determine water policy and to define and clarify how
water will be used. Not only is this system difficult
to change, but it offers few incentives for conserving
water as exemplified by the "use or lose" and "nonimpairment" doctrines.

Jurisdictional Disputes

The uncertainty and complexity of the water policy
system along with the growth in water demands have
created conflicts over political autonomy, authority,
and responsibility among governmental units and between
public and private sector participants. Jurisdictional
disputes occur within states, such as conflicts between
west and east slope interests over the diversion of
Colorado River water to Denver; and they occur between
states, such as conflicts between states of the Lower
and Upper basins of the Colorado over which states have
responsibility for meeting the 1.5 million AFY obliga-
tion to Mexico. However, the most significant jurisdictional issue concerns individual state responsibility
versus more centralized federal responsibilities. Recent
federal-state conflicts include proposals which would:

• Limit the amount of land a farmer is allowed
  to irrigate with water from federal reclamation projects;

• Require states to pay a percentage of the
  front-end costs of reclamation projects; and

• Establish a national water policy with
  strengthened federal controls on western water
  resources.

These disputes are indicative of the increasing seriousness of water resource problems, including the question
of the extent to which western water should be used to
meet national energy needs.

Reserved Water Rights

Reserved rights recognize that when the U.S. establishes a federal reservation such as a national park, military installation, or Indian reservation, a sufficient quantity of water is "reserved" to accomplish the purposes for that particular land. This reserved rights doctrine is significant because the federal government and Indian tribes own large amounts of western land--about 70 percent of the land in the Colorado River Basin (CRB). Thus, large quantities of water are at stake. At the present time relatively little water has been put to use under this doctrine, but these rights are not subject to state appropriation law and may not be lost if they are not used. If Indians and federal agencies do exercise these rights for large quantities of water, then existing users with appropriated rights could be affected. This situation adds uncertainty about the amount of water available to existing and future users under state appropriation systems.

Pollution from Energy Facilities

Current federal and state regulations are designed in essence to prohibit the direct discharge of pollutants from energy facilities into surface streams. Nevertheless, even if these standards are met, energy production and conversion processes can pollute both surface and groundwater either through the disposal of waste products or through the disruption and contamination of aquifers during mining and in situ operations. For aquifer contamination from mining and in situ recovery, scientific information on the extent of the potential impacts and possible control technologies is inadequate. Control of seepage and runoff from waste disposal sites constitutes a potentially serious problem, especially after a facility has shut down. Given current federal and state regulations and the level of scientific uncertainty, existing environmental regulations are inadequate to ensure that long-term or irreversible damage to surface and groundwater quality does not occur.

Salinity Control

Salinity has already been singled out for regulatory control by the federal government and by each of the states in the CRB and is of increasing concern in the Yellowstone River Basin. The major sources of salinity are natural salt flows and runoff from irrigated agriculture. The increase in salinity concentration from energy development is expected to be small relative to existing levels. Nevertheless, energy development is likely to intensify conflicts over salinity control, and thus energy production could be constrained by salinity standards if adequate controls are not established. Federal efforts to control salinity in the CRB have already been subject to delays and cost increases, and states of the CRB have been criticized by some groups for failure to establish a comprehensive, authoritative, and enforceable salinity control program.

Municipal Wastewater Pollution

Rapid, relatively large population fluctuations in small western communities near energy development projects can cause water pollution due to inadequate treatment of municipal sewage. Without increased assistance, few small communities in the West will be able to afford the cost of upgrading capacities to meet demands or of installing secondary and tertiary treatment required by the Clean Water Act (1977).

POLICY RESPONSES

Many alternatives exist for dealing with these problems and issues. This study has identified and evaluated several choices for conserving water, augmenting water, protecting water quality, and improving the capacity to manage the resource. The overall purpose of this analysis is to inform policymakers about the range of trade-offs which can be anticipated if various choices are made. Inclusion of an alternative does not imply that it is a reasonable choice for resolving the issues; but many of the options discussed have received attention or have been suggested elsewhere.

In the previous five chapters, the policy options were organized as follows:

Chapter 4: Conservation of Water

Chapter 5: Augmentation of Water Supply

Chapter 6: Water Quality Protection

Chapter 7: State Water Administration and Management

Chapter 8: Regional and Federal Roles in Water Management

In this chapter, major findings from these five chapters will be summarized using the following categories:

- Federal and State Responsibilities
    - State Government Options
    - The Federal Role
- Mitigating Water Shortages
    - Augmentation vs. Conservation
    - Water Conservation in Energy Facilities
    - Saline Water Use by Energy Facilities
- Protecting Water Quality

Federal and State Responsibilities

One clear implication of our analysis of problems and issues is that better management strategies are needed. Among the most important needs in this regard are:

(1) Predevelopment monitoring of groundwater resources and quality to improve knowledge about the degree of protection required and the effectiveness of control technologies;

(2) Enhanced coordination of water planning and regulation, particularly to integrate water availability and water quality concerns; and

(3) Improved efforts to address multistate problems, particularly salinity control and water availability within basins and sub-basins.

A central question in dealing with these and other management issues is how the responsibility will be shared among state governments, regional organizations, and federal agencies. Water resource management in the West has been primarily a state responsibility. Certainly states do not want to lose this authority and control, since water issues are central to questions of future economic growth and development in the region. As a result, concerns about water often are raised by state officials as much to resist any changes which would increase federal control as to try to resolve specific issues related to water availability or water quality.

However, states are facing increasing pressures for a modified federal-state relationship. These pressures are related to:

- A perceived need for states to improve their own water management capabilities;

- The high percentage of federal land ownership in the region;

- National interest in preserving the West's scenic and recreational areas;

- The fact that a large number of the West's water projects result directly from federal financing; and

- National energy goals which could lead to large-scale increases in the development of western energy resources.

Two recent examples of these pressures for change include the national water policy initiative and the declaration of federal reserved rights (see Chapter 2).

Chapters 7 and 8 evaluate several policy alternatives for improving state, regional, and federal water management. These alternatives are discussed in terms of the following criteria:

- Will the information base on water uses be improved?

- Will state flexibility in meeting multiple demands for water be increased?

- Will water quality and water quantity concerns be integrated?

- Is the economic efficiency of water use improved?

- Are social and environmental factors considered? and

- How feasible is the alternative?

While management of western water resources will continue to require a mix of state and federal responsibilities, our analysis leads us to conclude that states should continue to be the level of government primarily responsible for managing water resources. This general conclusion can be most directly attributed to the diversity of values and problems existing across the eight states in our study area, and the fact that water resources are directly related to state priorities for growth and development. Centralized or uniform approaches are not appropriate to managing these diverse situations. In addition, federal interests in the region are very diverse and complex: they include encouraging energy development; preserving wilderness, scenic and recreation areas; protecting wildlife; subsidizing agricultural development; and regulating water quality. Thus, a centralized federal policy would require, at a minimum, both a better consensus about priorities and a more coherent set of federal policies. This conclusion does not suggest that federal policies and programs cannot be improved or that the federal-state relationship cannot be strengthened. In fact, we believe both are critical to effectively managing future water problems.

State Government Options. It is clearly in the interest of state governments to improve water management and administration, both to preempt federal intervention and to better manage this critical resource. As indicated in Table 9-2, four policy options to improve state management flexibility and capacity have been evaluated.

The most important difference among these options is whether responsibility for dealing with changing demand patterns will rest primarily with state governments or with the private sector. Coordinated administrative systems and publicly-operated water banks would provide the greatest degree of state control. In contrast, reducing institutional barriers to water transfers would gradually allow the market system to play a bigger role in how water is used. Removal of at least some of the institutional constraints appears to be a prerequisite for water rights exchange systems.

Economic effects also can be expected to differ substantially among these choices. Since energy interests will be able to pay a great deal more than agriculture for water rights and water transportation systems, pressures will increase for states to facilitate water transfer. Water banks facilitate such transfers most directly by establishing a mechanism for coordinating transfers. Such transfers could be very beneficial economically to individual farmers, but costs would include

TABLE 9-2:   Policy Options for State Government
             (Chapter 7)

Policy Option	Description
Reduction of Institutional Constraints to Transfers	To meet changing needs for water, modifications in the legal/administrative system could include (1) broadening the definition of "beneficial use"; (2) establishing administrative procedures to settle disputes over impairment; (3) simplifying the process for separating water rights from land ownership; and (4) allowing states to control the transfer of water from federal projects.
Water Rights Exchange Systems (e.g., water banks)	Water banks, an explicit mechanism for transferring water rights, could be established on a state or regional level. Their purpose is to coordinate demands for water with availability. They could be established privately as a cooperative enterprise among irrigation districts or publicly as part of state water regulatory agencies.
Time-Limited Permits	New permits for water rights would be granted for specified time limits, after which the water rights would revert back to the states. They could be applied inclusively to all new water rights requests or selectively to particular user groups, such as industry.
Coordinated Administrative Structures	Such options are intended to increase the responsibility for day-to-day management of water and to integrate water quality and water quantity concerns. One example is the so-called "Utah Plan" which would organize water planning around local areas or hydrological units.

loss of income to the agricultural sector as a whole, reduced agricultural production, and the possibility of water rights speculation which would increase water prices.

All of these options can also improve state management by improving the information base about water resources. The "Utah Plan," an example of coordinated administrative systems, most directly addresses information needs by integrating water quality and availability information around hydrologic units. A water rights exchange bank could serve as a clearinghouse for information on water quality and availability.

Only one of the alternatives discussed in this chapter, coordinated administrative systems, is specifically designed to address the problem of integrating water quality and quantity decisions--a particular need presently in the CRB. In the "Utah Plan," this would be accomplished at the local level. Coordinated administrative systems could take a variety of forms, including establishment of departments of water resources which would be mandated to balance availability and quality concerns.

Feasibility is a realistic concern of almost any change in existing administrative systems. This is in part because some of these options would make significant changes in laws, doctrines, or state responsibilities. Most of the alternatives would probably be opposed by agricultural interests. However, it is also true that many of these options have already been introduced in some states. Laws or administrative policies regarding beneficial use have been changed to account for instream values in Colorado, Montana, New Mexico, and Arizona, and to allow transfers in Idaho. Water banks were established in Idaho and were used successfully in California in the 1976-77 drought. Time-limited permits have been established in Florida, Iowa, and Utah. Because of increasing demands for water and pressures to improve management of the resources, these approaches to water administration are likely to become more feasible over the next several decades.

The Federal Role. Both federal agencies and state governments have legitimate interests in western water resource management and, as the discussion of problems and issues suggests, a clear need exists for improved regional management of water availability and water quality (e.g., salinity control). Three alternative federal roles to help meet this need have been evaluated (see Table 9-3).

TABLE 9-3:   Policy Options for Regional and Federal
             Water Management
             (Chapter 8)

Policy Option	Description
Technical Assistance	Two forms for providing federal research and development assistance to the states to improve management capacities are: (1) baseline information on water use, availability and quality; and (2) coordinated grant support to assist states in developing information for administering water resources.
Changing Federal Policies	Quantifying reserved rights by administrative adjudication, as opposed to the current practice of judicial adjudication, attempts to reduce the uncertainty of future water availability. "Real-cost" water pricing bases the price of using water from federal projects on costs rather than the ability of the user to pay.
Quantification of reserved rights	
"Real-cost" water pricing	
Federal-Interstate Compacts	To provide comprehensive river basin planning and management, a commission consisting of the governor (or designee) of each state and a federal member appointed by the president would be established. The commission would develop plans, adopt and promote uniform and coordinated policies, and encourage development of projects in accordance with the plans.

Among these approaches, federal technical assistance to states in developing water management information systems could be the most effective option for improving state and interstate water management. Although a wide range of water data is already collected by federal and state agencies, two major deficiencies exist. First, the data are typically isolated to the missions of one federal agency rather than to specific needs of states; thus, research and development programs need to be targeted to state water management agencies. Second, existing data are limited in scope; for example, water availability research efforts are typically limited to the "supply side," or stream flow measurements, rather than the "use side." Among the most critical information needs are return flow measurements, actual use patterns, and water quality monitoring, particularly regarding predevelopment conditions and the effectiveness of holding ponds (see Chapter 6).

Because water appropriations are the responsibility of state governments, information collection and management also should be largely state responsibilities. However, federal assistance or incentives are advantageous for three reasons. First, costs are more affordable at the federal level. Second, implementation would be facilitated by the existence of several programs already channeling federal funds to the states. For example, the Office of Water Research and Technology (OWRT) provides $120,000 per year to research centers in each state to improve water management. The mission of OWRT could be expanded to meet future management needs without requiring new administrative systems. Third, technical assistance is more feasible than many other federal activities since it does not intervene directly in state water policymaking; rather, it is intended to improve state management capacity.

Other options considered in this report for an expanded federal role are less attractive for a variety of reasons. Although quantification of reserved rights and "real-cost" water pricing have received considerable recent attention, both face strong barriers. There is little evidence that an acceptable criterion for quantification can be found which will satisfy states, Indians, and the federal government. Without such a criterion, or at least a basic agreement among the parties-at-interest, quantification is not likely to reduce the delays and uncertainties associated with litigation over water rights. If an acceptable criterion were found, administrative adjudication is a relatively long-term (10 to 15 year) process, although the process may be quicker than the current judicial adjudication procedures.

"Real-cost" pricing for reclamation water addresses a very limited goal of forcing future users of water

from federal projects to pay the full cost of making
water available. The limitation to future water users
would mean that the impact in well-developed river
basins such as the CRB would not be great. While "real-
cost" pricing may encourage some agricultural users to
irrigate more efficiently, its more likely effect will
be to promote nonagricultural uses of future federal
project water. If this occurs, energy developers are
the likely beneficiaries.

Federal-state cooperation in regional water management is paradoxical. It is critically needed in the CRB
to deal with such problems as salinity control and
equitable water allocation. Yet, the feasibility of
options such as a federal-interstate compact under present conditions is remote. This is partially because of
the increasing tension between federal and state governments over water, but it is also because of the strong
tendency for independence in each of the states. Thus,
it is not likely that the necessary agreement can be
reached to form a compact, particularly for the seven
states of the CRB.

Nevertheless, if the implementation barriers could
be overcome, substantial advantages could be obtained by
such a regional organization. The National Water Commission advocated the federal interstate compact as the
preferred mechanism for water resource planning and
management in multistate regions (1973, p. 424). Since
few water problems conform to state boundaries, its primary advantage is in providing a basin-wide capacity for
integrating water quality and water availability concerns among states of the basin. While obviously requiring that states give up some authority, it would
strengthen regional responsibility. It would not necessarily reduce states' responsibility for "day-to-day"
management, but state policy would have to conform to an
overall regional management plan.

Mitigating Water Shortages

A variety of policy alternatives has been discussed
to help mitigate potential water shortages and the disputes arising over how scarce resources should be used.
These options are summarized in Table 9-4. Among these
alternatives, two specifically address water requirements for energy development: water conservation in energy facilities and use of saline water. Each of these
alternatives has several advantages and is discussed in
more detail in succeeding sections. First, however, the
following section will summarize the conclusions regarding the other conservation and augmentation options considered.

TABLE 9-4: Policy Options for Conservation and
Augmentation

Category	Specific Alternative	Description
Conservation (Chapter 4)	Conservation in energy facilities	Two basic approaches: (1) modifying the process design in synthetic fuel processes to maximize water reuse, and (2) using "dry" instead of "wet" cooling towers (or some mix of the two)
	Improved agricultural water use	Improvement in off-farm delivery systems (such as canal lining, removal of noncrop vegetation, and flow regulating systems), and improvements in on-farm irrigation practices (such as ditch lining and sprinkler irrigation)
	Crop switching	Replacing water-intensive crops (such as alfalfa and other forage crops) with vegetables or wheats
Augmentation (Chapter 5)	Use of saline water in energy	Using saline water (i.e., water with a TDS content of at least 1,500 mg/l), which is not acceptable for many other uses
	Groundwater use and storage	Conjunctive use of surface waters and groundwater, mining of deep aquifers, and storage in aquifers
	Weather modification	Increased precipitation from natural storm systems, primarily increased snowfall (i.e., snowpack augmentation)
	Vegetation management	Removing natural vegetation to increase upland runoff or reduce water loss by evapotranspiration
	Surface water storage, transfer, and diversion	Reservoirs to regulate the timing of water availability, transfer between river basins (interbasin transfers), and intrabasin transfers

mg/l = milligrams per liter

Conservation vs. Augmentation. Generally, augmentation strategies are very costly ways to deal with the complex problems and issues affecting western water resources. Substantial quantities of water can be added by some of these alternatives but, when considering economic costs, environmental damage, and long-term ecological risks, most of these choices are likely to worsen rather than improve conflicts over the appropriate use of water. Previously, the approach to western water problems largely emphasized technological fixes to get water where it was needed. Although such approaches have proven valuable in the past, they are limited as solutions to future problems. Solving future problems will require a mix of approaches, including some of the augmentation choices discussed in this report. For example, groundwater use has been, and will continue to be, an important option in certain localized cases. However, a mix of alternatives will increasingly need to offer options other than augmentation, including conservation and improved management of the resource.

Augmentation alternatives do have some advantages. Groundwater use, either separately or conjunctively with surface water, can provide a short-term, localized benefit, particularly for energy facilities. However, in some areas groundwater is already being used by agriculture and municipalities at rates beyond annual replenishment; therefore, these sources would probably be unavailable to meet energy needs. There are deep aquifers underlying shale and coal resources which could be developed rapidly for energy conversion facilities. Further study would be needed on a site-by-site basis to provide information on the ability of these aquifers to sustain withdrawal rates during the twenty- to thirty-year life of the commercial units.

Both weather modification and vegetation management can increase water supplies by one to two million AFY in the CRB. Although economic costs for these options appear to be very low, they both may have high environmental costs and risks. Studies would be needed to assess potential ecological damage. Also, the time span required to have fully operational programs further reduces the effectiveness of these options as a short- or mid-term solution. Important legal questions concerning the rights to the increased water supply exist for both options. New legislation may be needed to assure that the benefits from weather modification or vegetation management accrue to those who produced the additional water supply.

Interbasin and intrabasin transfers and increased storage have increased usable water supplies substantially in the CRB. However, their future use is limited by several factors. First, the Colorado River is the most regulated river in the world; thus, few acceptable locations for large storage projects remain. Second, there is a penalty for using impoundments since water is lost to evaporation. In the CRB, about 14 percent of the annual average flow (about two million AF) is lost to evaporation. Third, substantial opposition exists to these projects on environmental grounds, particularly in the Yellowstone River which is free-flowing. Finally, interbasin transfers to the CRB have been prohibited for the past twelve years by the Colorado River Basin Project Act (1968), which prohibits the Department of Interior from carrying out "reconnaissance" (investigative) studies of this option.

As is true of many augmentation choices, improved irrigation efficiency and crop switching have serious disadvantages for dealing with water shortages. Considerable uncertainty exists about how much water can be "saved" by irrigation improvements because of questions about the amount and nature of irrecoverable losses in any given irrigation project. Nevertheless, it is generally agreed that the water "saving" potential is not as significant as the potential for improved water quality. In fact, current salinity control efforts in the CRB focus largely on improvements in irrigation practices. Also, by reducing the amount diverted, instream flow values will be protected, but this will need to be balanced against the loss of wildlife habitat dependent on agricultural runoff. Although switching to less water-intensive crops could save large quantities of water, this is not a feasible region-wide policy option, given the climate across the area and the current makeup of the agricultural economy.

Improvements in irrigation efficiency will face both economic and institutional obstacles. Generally, the economic costs far exceed the benefits to the individual farmer. If this option is to be implemented on a large scale, it will probably require additional federal subsidies. In addition, even if farmers conserve water, the "use or lose" doctrine means that water "saved" by the individual farmer cannot be retained.

Water Conservation in Energy Conversion Facilities. Process design decisions in coal synfuel facilities help to determine the fresh water requirements, the amount of contaminated water which must be treated, and the amount of water which must be disposed. Two examples of process design changes were analyzed: the Western Gasification Company Lurgi gasification plant in northwestern New Mexico and the Exxon liquefaction process in Wyoming. Water minimization can be achieved through maximizing the reuse of wastewater from one process to supply another (for example, by recovering exhaust steam from turbines and reusing it for steam generation) or through changes in the conversion process (for example, by changes in the method of producing hydrogen).

In addition to reducing raw water requirements, a second major advantage of process design changes is the reduction in wastewater disposal requirements (see Chapter 4). For example, modifications of the Exxon process decreased the plant water treating load by approximately one-half of the base case rate. Existing knowledge suggests that process design changes will either save money or only slightly increase costs. A wide range of possible process design changes exists; thus, the economic effects will depend on a variety of site-specific factors, such as raw water cost. However, Exxon believes that "the economics of water optimization will have only a minor effect on the selling price of coal liquids and at this stage of the overall process development, the effect can be disregarded" (Exxon 1977, p. 99).

A second approach for conserving water in energy facilities is the use of dry cooling. Since water for cooling is the largest single water requirement for electric power plants and synthetic fuel facilities, the choice of cooling technology is crucial. Table 9-5 summarizes the findings on water savings. Wet/dry cooling of power plants can reduce water requirements by about 70 percent, while intermediate wet cooling of coal synfuel plants can reduce water requirements by about 20 to 30 percent. These water savings can increase economic costs. In the case of synfuel plants, however, the economic penalty for intermediate wet cooling is expected to be less than 0.5 percent. For power plants the economic penalty would typically be from 3 to 5 percent. In addition, with all dry cooling of power plants, plant efficiency and capacity could be lowered due to larger energy requirements to run the fans and due to higher condenser temperatures on hot summer days. For this reason, wet/dry cooling is generally much more practical than all dry cooling for power plants.

Although the economics are not always favorable, conservation techniques in energy facilities may be attractive for other reasons. In some water-short areas, water supplies may be physically or institutionally limited; thus, water conserving technologies may be necessary. And, even if sufficient water is available, a facility with smaller water requirements will be less threatening to other water users. This factor is likely to become increasingly important in the siting process.

TABLE 9-5: Water Savings for Coal Conversion Facilities
           With Alternative Cooling Systems[a]
           (AFY)

| Plant (size) | Water Requirements | | |
	High Wet Cooling	Intermediate Wet Cooling	Minimum Wet Cooling
Power Plant (3,000 MWe)	23,900-29,800	5,500-9,500	NC
Lurgi (250 MMscfd)	4,900-7,100	3,300-5,600	2,900-5,200
Synthane (250 MMscfd)	7,700-8,700	5,900-6,700	5,500-6,300
Synthoil (100,000 bbl/day)	9,200-11,800	7,500-9,700	7,000-9,100

NC = not calculated

[a]Range is across the sites considered (see Chapter 4).

Water conservation for energy resource development
may be the easiest set of alternatives to implement.
Conservation for energy resource development can save
large percentages of water and has less uncertainty
associated with its effectiveness and cost. It appears
easier to implement new regulations on energy resource
development than on agriculture, and energy industries
are probably better able than agriculture to afford the
cost increases.

Saline Water Use by Energy Facilities. Saline water
(containing more than 1,500 parts per million of TDS) is
a resource not suitable for use by many of the competi-
tors for western water, particularly agriculture and environmental interests. However, energy developers can
afford to treat saline water at only modest cost increases. Several sources exist: (1) surface waters
such as the Powder River of the Yellowstone River Basin,
(2) aquifers such as the Madison or Dakota, (3) irrigation return flows from irrigation projects, and (4)
municipal wastewater.

The economic costs of using saline water include supply, treatment, and disposal of brines. As shown in Table 9-6, total cost increases from using saline water will be modest: a maximum of 6 percent for gasification, 3 percent for electric power generation, and 2 percent for liquefaction. Since no credit has been given for the nominal treatment, supply, and disposal costs for facilities handling fresh water, these numbers are worst case estimates.

Several incentives exist for energy developers to use saline water. These include the possibility that fresh water supplies will be unavailable or limited; the possibility that political conflicts, lengthy adjudication proceedings, and other siting delays could be reduced by using this "unwanted" water source; and the need to have alternative water sources as a hedge against the uncertainties of the water situation in the West. States and federal agencies also encourage saline water use, for example, as part of a salinity control program or to facilitate energy development. State policies in this regard could include allowance of rate adjustments (so that increased costs are borne by consumers) and financial subsidies such as deferred or reduced taxes.

TABLE 9-6: Estimated Costs of Using Saline Water

Category	Gasification ($/Mcf)	Liquefaction ($/bbl)	Electric Power Plant (¢/kWh)
Boiler	0.01	0.10	0.007
Process	0.009	0.06	Neg.
Cooling Towers	0.03	0.03	0.05
Disposal	0.16	0.52	0.014
Water Supply	0.06	0.22	0.044
Total Costs	0.27	0.93	0.11
Approximate Product Price	4.50	40.0	4.0
Cost as Percent of Price	6%	2%	3%

Mcf = thousand cubic feet   kWh = kilowatt-hour

Source: See Chapter 5

Protecting Water Quality

Three general water quality problems associated with energy development have been identified above: pollution from energy facilities, salinity control, and municipal wastewater treatment. Since water use and water quality are directly related, policies designed for water availability questions can also be important to water quality. For example, improved irrigation efficiency can provide some water savings and major water quality benefits. Four additional policy options specifically addressing water quality were assessed, as indicated in Table 9-7.

The option to improve water quality controls for energy development does not represent a radically new program, but rather an expanded and more comprehensive approach to ongoing activities. The goals of this option are to ensure that best available practices are used in the short term and that over the longer term the knowledge about geohydrology and the behavior of potential contaminants is improved. This will ultimately improve the ability to define the risks of energy development and the effectiveness of various control measures. Without such a program, the probability increases that serious and potentially irreversible damage will occur to the West's water resources, especially groundwater. The disadvantages of this option include increased energy costs due to stricter control measures and increased public sector expenditures for research. However, given the high level of uncertainty concerning potential impacts and the likelihood of expanded production of western energy resources, the costs appear to be worth paying.

Temporary sewage treatment would allow the use of sewage lagoons or waste stabilization ponds in energy-impacted communities until more conventional secondary treatment facilities could be constructed. The primary advantages of this approach are that capital and operating costs would be lowered and the time needed to bring the system on-line shortened. Thus, sewage lagoons provide a flexible, economical means of providing municipal wastewater treatment capacity until the population stabilizes and the necessary front-end financing from the expanded tax base is available. The major disadvantage is the inability to consistently meet EPA's secondary treatment standards. While this problem could be dealt with by complete retention of wastes, this would also increase consumptive water use.

Salinity problems are already an important issue in the CRB and some streams in the Yellowstone River Basin. Although the primary sources of salinity are agriculture and natural salt flows, energy development will increase the problem through salt loading and the downstream

TABLE 9-7:   Policy Options for Protecting Water Quality

Policy Option	Description
Improved Water Quality Controls	Improving water quality planning and information through: (1) predevelopment monitoring of groundwater quality; (2) control plans for each energy facility; and (3) monitoring and research programs
Temporary Wastewater Treatment Measures	Allows use of sewage lagoons (or waste stabilization ponds) as a temporary measure until more conventional secondary treatment facilities can be constructed
Salinity Control Projects	Approach being emphasized in CRB. Includes desalination plants, diverting flows around areas of high salt pick-up, evaporating saline sources, and improved irrigation practices
Salinity Offset Policy	Proposed energy developments which increase salinity allowed to proceed only if some offsetting activity were implemented so that salinity concentrations would not be increased over some specified time period

concentrating effects of consumptive use.   Construction of salinity control projects is the approach currently emphasized in the CRB.   The other choice discussed in this report is a salinity offset policy specifically defined to deal with salinity increases caused by energy development.   Both options could provide important environmental and economic benefits.   The economic benefits for agriculture, municipalities, and industry for each mg/l decrease are estimated at $343,000 per year.

Because constructing salinity control projects is a comprehensive alternative, it can have a great effect on salinity levels.   For example, three projects authorized by the 1974 CRB Salinity Control Act (Grand Valley, Paradox Valley, and Las Vegas Wash) are estimated to decrease salinity concentrations at Imperial Dam by about 43, 18, and 9 mg/l, respectively.   The disadvantages are that large federal capital outlays are required and water availability is decreased.   Cost estimates for these three projects range from $0.2 to $1.0 million per mg/l salinity reduction at Imperial Dam.   Although opposition to this approach exists and the authorized projects have been subject to considerable delays, the legal and institutional basis already exists in the CRB through the CRB Salinity Control Act and the CRB Salinity Control Forum.

The salinity offset policy would establish an administrative mechanism to prevent increased salinity levels due to energy development. The primary advantage to this approach is economic efficiency, since energy developers could use the lowest cost "offsets" available. Increased energy costs are expected to be small. If offsets were obtained by paying a proportion of the costs for the salinity control projects discussed above, energy costs for three facilities analyzed are estimated to increase by 0.02 to 0.5 percent. Two important disadvantages exist for this policy. First, the salinity models required to predict salinity impacts are not well developed. Second, by adding another requirement to energy facility siting permits, energy development could be slowed or constrained. However, states could help to alleviate this problem, for example, by banking offsets. The legal basis and practical feasibility of such a policy is uncertain. One type of salinity offset policy has been attempted in Colorado by a local government agency but the plan was vetoed by Governor Lamm. This plan would have affected a number of future water users in the state and was therefore strongly opposed by several groups. By limiting the offset policy to energy development activities the opposition would be reduced, although important equity questions would remain.

A SUMMARY COMPARISON WITH OTHER STUDIES

This is one of several recent studies[1] which address the relationship between water resources and energy development in the West. Since the findings of these reports differ in some significant ways, it is useful to compare them. This section presents a brief comparison of this study with two other works--one by the Colorado Department of Natural Resources (DNR) (1979, for the Water Resources Council) and one by the General Accounting Office (GAO) (1980). Table 9-8 summarizes the nature and findings of these three studies. As indicated in the table, some differences exist in their purposes, scope, and approaches. However, there are enough similarities so that it is possible to compare the findings and conclusions.

Similarities among the studies include the following: (1) considerable overlap exists in the study areas (although the Colorado DNR study is limited to the Upper CRB); (2) all three studies emphasize coal and oil shale development between the present and the year 2000;

(3) the general purpose of each study is to address potential water constraints on energy and the impacts of energy development on water resources and current water users; and (4) the approaches taken are similar in that each study assesses water requirements for individual energy technologies, estimates quantities of water which will be required for various levels of energy development, and compares this with water availability estimates. Among the most significant differences in the purposes and approach taken in these studies are the following: (1) the GAO and Colorado DNR studies place emphasis on the question, "Does enough water physically exist to support western energy development?", while our study emphasizes the water availability issues associated with energy development; (2) our study is more directly concerned with the water quality impacts of energy development; and (3) while the GAO and Colorado DNR studies do address some of the institutional factors that affect the availability of water, these factors are given more emphasis in our study.

As summarized in Table 9-8, different conclusions are reached. Both the Colorado DNR and the GAO reports are relatively optimistic that sufficient water exists to support energy development without harming other uses. Our study concludes that water availability and quality issues will continue to be major factors influencing the type, magnitude, and location of development. While these findings do represent important differences among the studies, they are not completely incompatible. There is general agreement regarding phy-sical availability of water resources and water requirements for energy. All three studies conclude that water requirements for energy will represent only a small percentage of total water availability on a basin-wide or regional level--even in the Upper CRB where water supplies are the tightest.

The fundamental differences in the conclusions and emphases seem to arise from differing judgments as to what roles social, institutional, and political factors will play in water/energy developments. Our judgment is that conflicts over water resources will be critical in determining the future use of water, as has been demonstrated, for example, by the difficulties slurry pipeline companies have had in obtaining water rights. The Colorado DNR study explicitly states that their conclusions only take into account those institutional factors embodied in each state's water rights system and in the "Law of the River" (e.g., international treaties, interstate compacts, and U.S. Supreme Court decrees). They note that other institutional factors are important (primarily federal environmental regulatory laws and programs), but it is impossible to quantify the effects such factors may have on the availability of water for

TABLE 9-8: A Summary Comparison of Three Studies

	Colorado DNR Study	GAO Study	Energy From the West
Purpose	To assess water requirements of energy technologies; water availability; the effects of development on hydrology and water quality; the economic costs of providing water and managing wastewater; and the social, economic, and environmental impacts of energy development (pp. xv-xvi).	To discuss the changing conditions that have reduced projected energy-related water demands and to discuss existing constraints on energy and mineral developers' access to federal project water (p. 1).	To identify water impacts and issues associated with western energy development and to evaluate policy alternatives for dealing with energy-related water availability and water quality issues (Chapter 1).
Scope			
Time	Present to the year 2000	Present to the year 2000	Present to the year 2000
Resource	Coal, oil shale	Coal, oil shale, uranium, and geothermal	Coal, oil shale, uranium, oil, natural gas, and geothermal
Area	Upper CRB: Arizona, Colorado, New Mexico, Utah, and Wyoming	Arizona, Colorado, Montana, New Mexico, North Dakota, Utah, and Wyoming	Arizona, Colorado, Montana, New Mexico, North Dakota, South Dakota, Utah, and Wyoming
Approach	Predictions of water availability are based on an assessment of existing conditions in the Upper CRB and several scenarios which estimate future water needs for conventional uses and for emerging energy technologies. Scenarios are based on data from each state regarding most probable levels of future development. Analysis includes detailed presentation of data, assumptions, and references.	Primary approach is to review and update previous assessments of water availability for energy. Conclusions are based on an assessment of previous studies and on personal contacts with unidentified officials from state and federal governments, representatives from energy companies, and consumers of water. Data sources and documentation are only occasionally provided.	This study is part of a larger technology assessment of western energy development. Impacts are identified largely from several alternative scenarios representing various alternative possibilities. Data sources are primarily existing reports and institutional analysis of legal, political, and social factors which affect water and energy policy.

TABLE 9-8: (Continued)

| Findings/ Conclusions | Water demands of emerging energy tech- nologies producing 1.5 million bbl/day could be satisfied from surface sup- plies without having to significantly (if at all) reduce other projected consumptive uses in the Upper Basin. This is premised on four qualifiers: (1) water not presently under contract is purchased for the Bureau of Reclama- tion reservoirs; (2) new storage and transport facilities are built; (3) minimum instream flows are not con- sidered; and (4) several institutional factors including interstate compact provisions will not restrict overall development (pp. 1-1 to 1-3). | Water requirements for energy are much less than previously assumed; thus, adequate water is available for energy development through at least the year 2000. Development is possible without interfering with existing users or proposed projects because water will be available from new storage facil- ities, from other rights holders, and from federal project water (pp. iv-v). Existing uncertainties will only limit the number of sites where development can occur. | Water availability and water quality issues must be considered together and they will continue to be major factors influencing the type, magnitude, and location of energy development in the West. Although enough physical sup- plies exist on a regional or basin-wide level to support large increases in de- velopment, considerable uncertainty ex- ists about water availability on a site- specific basis. Institutional, social, and environmental factors are at least as important as total physical supplies in determining water availability for energy. A variety of policy options exists for dealing with these issues. |

energy or other uses (Colorado DNR 1979, p. 1-2). The GAO study acknowledges that uncertainties exist, including environmental requirements, reserved water, and instream flows, but concludes that they will influence only site selection and not overall development levels (1980, p. iii).

Based on our own study and the ones by GAO, the Colorado DNR, and several others, two general conclusions can be drawn as a guide to future research in this area. First, research efforts need to address explicitly the relationship of water quality and water quantity. While this includes a wide range of interrelationships, among the most important concerns are: (1) the impacts of withdrawals on water quality; (2) the water quality consequences of mitigating water availability problems; and (3) the institutional mechanisms appropriate for managing quality and quantity problems together. Second, it is unreasonable to expect simple quantitative answers to the question: "Will water availability be a constraint on western energy development?" Such a question cannot be answered by just comparing water needs versus supply. Strictly quantitative analyses are subject to considerable uncertainties--related to data limitations, changing events, and policy priorities. Additional research should focus on western water resource issues and how these issues can be dealt with. Such knowledge bases need to include both technical considerations, such as process design options for water minimization, and institutional improvements, such as enhanced basin-wide management.

Regarding policy directions, the central conclusion of our work is that the western water situation has changed rapidly as multiple demands for a scarce resource continue to increase. These demands include energy development, national defense uses, mineral and industrial expansion, reserved rights, instream values, salinity control, and population growth. The implication of this growth is that the water policy system, which has dealt successfully with many water problems in the past, must adapt to these changing circumstances in order to balance the competing needs for water in the future. A variety of proposals has been made to correct the deficiencies in the existing water policy system. In this regard, two general conclusions result from our study. First, water resource management should continue to be primarily a state responsibility; while several federal policies can be used to help the states, there is little evidence suggesting that a centralized or uniform federal approach to water policymaking is appropriate. Second, if states are to be successful in resolving water resource conflicts, three policy needs can be identified as guideposts to revisions in state policy:

- The need to encourage <u>flexibility</u> of water policies so that the <u>range of</u> legitimate, multiple uses of water is recognized;

- The need to encourage water resource <u>management</u> which enhances day-to-day <u>control</u> of the resource and integration of water quantity and quality concerns;

- The need to encourage public education and <u>participation</u> in water resource policymaking.

While these guidelines represent dramatic departures from traditional water policymaking in the West, they consistently emerge as requirements for dealing with the changes occurring throughout the West. If these guidelines are recognized, several augmentation, conservation, and management alternatives exist which can avoid or reduce many of the potential conflicts.

NOTES

[1]Among the more recent studies on this subject are: Andersen and Keith 1977; Boris and Krutilla 1980; Colorado, Dept. of Natural Resources 1979; U.S., Congress, Office of Technology Assessment 1980; U.S., Congress, Senate Com. on Energy and Natural Resources 1978; U.S., GAO, Comptroller General 1979, 1980; and U.S., Water Resources Council 1978.

Bibliography

Andersen, Jay C., and John E. Keith. 1977. "Energy and
the Colorado River." Natural Resources Journal 17
(April):157-68.

Arizona v. California, 373 U.S. 546 (1963), Decree 376
U.S. 340 (1964).

Ashland Oil, Inc., and Occidental Oil Shale, Inc. June
1977. Supplemental Material to Modified Detailed
Development Plan for Oil Shale Tract C-b, prepared
for Area Oil Shale Supervisor.

Atomic Energy Act of 1954, Pub. L. 83-703, 68 Stat. 919,
as amended by Pub. L. 91-560, 84 Stat. 1472.

Balzhiser, Richard E. 1977. "R&D Status Report."
EPRI Journal 2 (December):38-43.

Barlow, A. C. 1963. "Industry Looks at Saline Ground
Water." Ground Water 1 (No. 1):25-27.

Belle Fourche River Compact of 1943, Pub. L. 78-236, 58
Stat. 94 (1944).

Betz Handbook of Industrial Water Conditioning, 6th ed.
1962. N.p.

Bishop, A. Bruce. 1977. "Impact of Energy Development
on Colorado River Water Quality." Natural Resources
Journal 17 (October):649-71.

Bishop, A. Bruce, et al. 1975. Water as a Factor in
Energy Resources Development. Springfield, Va.:
National Technical Information Service.

Boris, Constance M., and John V. Krutilla. 1980. Water Rights and Energy Development in the Yellowstone River Basin: An Integrated Analysis. Baltimore: The Johns Hopkins Press (for Resources for the Future).

Boulder Canyon Project Act (1928), Pub. L. 70-642, 45 Stat. 1057.

Brown, T. C., et al. 1974. Chaparral Conversion Potential in Arizona, Part II, Forest Service Research Paper RM-127. Fort Collins, Colo.: U.S., Department of Agriculture, Forest Service, Rocky Mountain Forest and Range Experiment Station.

California, Public Utilities Commission, Utilities Division, Environmental Impact Branch. 1976. Early Draft EIS: Sundesert Nuclear Project, San Diego Gas & Electric Company, Application No. 53534. San Francisco: California, Public Utilities Commission.

California v. U.S., 438 U.S. 645 (1978).

Cazalet, Edward, et al. 1976. A Western Regional Energy Development Study: Economics, Final Report, 2 vols. Menlo Park, Calif.: Stanford Research Institute.

Chartock, Michael A., et al. 1981. "Environmental Issues of Synthetic Transportation Fuels From Coal," Final Summary Report to Energy Office, Office of Technology Assessment, U.S. Congress. Norman: University of Oklahoma, Science and Public Policy Program.

Choitz, Jackie. 1978. "Mining Law Update." Colorado/Business 5 (February):27-28.

Clean Water Act of 1977, Pub. L. 95-217, 91 Stat. 1566 (codified at 33 U.S.C.A. §§1251 et seq.)

Code of Federal Regulations (C.F.R.), Title 30, §816, as promulgated in Federal Register 44 (March 13, 1979):15395-.

Code of Federal Regulations (C.F.R.), Title 30, §822, as promulgated in Federal Resgister 44 (March 13, 1979):15450.

Code of Federal Regulations (C.F.R.), Title 40, §§122 and 146, as promulgated in Federal Register 45 (June 24, 1980):42472-512.

Colorado, Department of Health, Water Quality Control
Division (WQCD). 1979. "Water Quality and Mining:
A Process to Identify and Control Water Pollution
from Past, Present and Future Mining Activities in
the State of Colorado. Denver: Colorado, Dept. of
Health, WQCD.

Colorado, Department of Natural Resources (DNR). 1979.
Upper Colorado River Basin 13(a) Assessment: The
Availability of Water for Oil Shale and Coal
Gasification Development in the Upper Colorado
River Basin, Summary Report, A Report to the U.S.
Water Resources Council (Public Review Draft).
Denver: Colorado, Department of Natural Resources.

Colorado River Basin (CRB) Project Act (1968), Pub. L.
90-537, 82 Stat. 885.

Colorado River Basin (CRB) Salinity Control Act of 1974,
Pub. L. 93-320, 88 Stat. 266 (codified at 43 U.S.C.A.
1571 et seq. [Supp. 1976]).

Colorado River Basin (CRB) Salinity Control Forum.
1978a. Proposed 1978 Revision, Water Quality
Standards for Salinity, Including Numeric Criteria
and Plan of Implementation for Salinity Control:
Colorado River System. Salt Lake City: CRB
Salinity Control Forum.

Colorado River Basin (CRB) Salinity Control Forum.
1978b. Supplement Including Modification to Proposed
1978 Revision: Water Quality Standards for Salinity
Control, August 1978--Colorado River System. Salt
Lake City: CRB Salinity Control Forum.

Colorado River Basin (CRB) Salinity Control Forum.
1980. "Policy for Use of Brackish and/or Saline Water for Industrial Purposes." Unpublished paper,
Salt Lake City, Utah, September 11.

Colorado River Basin (CRB) Salinity Control Forum, Work
Group. 1975. Proposed Water Quality Standards for
Salinity: Colorado River System. Salt Lake City:
CRB Salinity Control Forum.

Colorado River Compact of 1922, 42 Stat. 171, 45 Stat.
1064, declared effective by Presidential
Proclamation, 46 Stat. 3000 (1928).

Colorado River Water Conservation District v. United
    States, No. 74-940, March 24, 1976, 44 L.W. 4372.

Comarc Design Systems; Wallace McHarg Roberts & Todd;
    and Northwest Colorado Council of Governments.
    1978. Draft Areawide Water Quality Management Plan
    for Eagle, Grand, Jackson, Pitkin, Routt and Summit
    Counties, Colorado. Denver: U.S., Environmental
    Protection Agency.

Committee for Economic Development (CED). 1976.
    Improving Productivity in State and Local
    Government. New York: CED.

Congressional Quarterly, Inc. 1973. Congress and the
    Nation, Vol. III: 1969-1972. Washington, D.C.:
    Congressional Quarterly.

Cook, C. Wayne, Colorado State University. September
    1979. Personal communication.

Costle, Douglas M. 1979. "Statement of Administrator,
    Environmental Protection Agency, Before the Sub-
    committee on Oversight and Review, Committee on
    Public Works and Transportation, U.S. House of
    Representatives," July 18.

Cotton, Gary D., San Diego Gas and Electric Company.
    November 19, 1979. Personal communication.

CQ Weekly. 1977. "Water Projects." July 2, p. 1377.

Crawford, K. W., et al. 1977. A Preliminary Assessment
    of the Environmental Impacts from Oil Shale
    Developments. Washington, D.C.: U.S.,
    Environmental Protection Agency.

Dames & Moore, Water Pollution Control Engineering
    Services. 1978. Construction Costs for Municipal
    Wastewater Conveyance Systems: 1973-1977.
    Washington, D.C.: U.S., Environmental Protection
    Agency, Office of Water Program Operations.

Davidson, Craig. 1977. "Mesita Battleground For Water
    Dispute." Denver Post, December 19.

Davis, Ray Jay. 1979. "Weather Modification, Stream
    Flow Augmentation, and the Law." In Rocky Mountain
    Mineral Law Institute: Proceedings of the 24th
    Annual Institute, July, 1978, pp. 833-63. New York:
    Matthew Bender.

Davis, Robert K., and Steve Hanke. 1971. Pricing and Efficiency in Water Resource Management, for the National Water Commission. Springfield, Va.: National Technical Information Service.

DeConcini, Dennis. 1977. Statement in U.S., Congress, Senate, Select Committee on Indian Affairs. Water for Five Central Arizona Indian Tribes for Farming Operations. Hearings, 95th Cong., 1st sess., May 23 and 24.

Delaware River Basin Compact (1961), Pub. L. 87-328, 75 Stat. 688.

Denver Federal Executive Board, Committee on Energy and Environment, Subcommittee to Expedite Energy Development, and Mountain Plains Federal Regional Council, Socioeconomic Impacts of Natural Resource Development Committee. 1975. A Listing of Proposed, Planned or Under Construction Energy Projects in Federal Region VIII, A Joint Report.

Denver Post. 1977a. "EDF to Sue for Water Salinity Control." April 15.

Denver Post. 1977b. "Limits on Irrigaton Opposed in Wyoming." November 16.

Denver Post. 1978. "Colorado Water Projects Endangered by Fish." April 20.

Denver Post. 1979. "Water Plan Given to Congress." May 17.

Denver Research Institute (DRI). 1979. Predicted Costs of Environmental Controls for a Commerical Oil Shale Industry, Working Draft Report. Denver: DRI.

Derthick, Martha. 1974. Between State and Nation. Washington, D.C.: The Brookings Institution.

Dewsnup, Richard L., and Dallin W. Jensen. 1975. Proposed Procedures for Planning, Allocating and Regulating Use of Water Resources in Utah, 2 vols. Salt Lake City: Utah, Division of Water Resources.

Dinwiddie, George A., et al. 1979. Plan of Study for the Northern Great Plains Regional Aquifer System Analysis in Parts of Montana, North Dakota, South Dakota, and Wyoming, Geological Survey Water-Resources Investigations 79-34. Lakewood, Colo.: U.S., Department of the Interior, Geological Survey, Water Resources Division.

Dollhopf, D. J., et al. 1977. Selective Placement of Coal Stripmine Overburden in Montana, II. Initial Field Demonstration, Research Report 125. Bozeman: Montana Agricultural Experiment Station, Reclamation Research Program, Montana State University.

Doneen, L. D. 1971. Irrigation Practice and Water Management. Rome: United Nations, Food and Agricultural Organization, Food and Water Development Division, Water Resources and Development Service.

El-Ashry, Mohamed T., and Robert M. Weaver. 1977. "Colorado Water Projects--Impacts and Alternatives." Denver Post, April 17.

Emergency Drought Relief Act of 1977 (Secretary of Interior, Emergency Action--1976-1977 Drought), Pub. L. 95-18, 91 Stat. 36, amended by Emergency Drought Relief Act (1977), Pub. L. 95-107, 91 Stat. 870.

Engdahl, Todd. 1979. "Colorado-Sun Belt 'War on River' Forecast." Denver Post, June 14, p. 1.

Environmental Defense Fund (EDF), Staff. July 1981. Personal communication.

Environmental Defense Fund (EDF), Staff. 1980. Personal communication.

Enzi, Michael B. 1977. "Energy Boom: Wyoming's Coal Veins Just Bring Troubles." Los Angeles Times, May 15, p. V-5.

EPRI Journal. 1978. "Disposal and Beyond." 3 (October):36-41.

Exxon Research and Engineering Co. 1977. EDS Coal Liquefaction Process Development, Phase IIIB, Quarterly Technical Progress Report for the Period July 1-September 30, 1977. Florham Park, N.J.: Exxon.

Federal Water Pollution Control Act Amendments of 1972, Pub. L. 92-500, 86 Stat. 816 (codified at 33 U.S.C.A. §§ 1251 et seg. [Supp. 1976]).

Fink, F. W. 1963. "Saline-Water Development Poses Problems in Metal Corrosion." Ground Water 1 (October):21-25.

Fischer, Roland (Chief Administrative Officer, Colorado River Water Conservation District). 1979. "Point of View: Water for Energy and Agriculture, Too." Denver Post, September 12, p. 21.

Fischer, W. H. 1975. "Weather Modification and the Right of Capture." Natural Resources Lawyer 8 (No. 4):639-58.

Florida Statutes, §§371.106 through 373.201.

Flug, Marshall, et al. 1977. The Impact of Energy Development on Water Resources in the Upper Colorado River Basin, Draft Report. Ft. Collins, Colo.: Colorado State University, Agricultural Engineering Department.

Gaufin, Sam O. 1977. "Colorado River Water Conservation District v. United States: An Increased Role for State Courts in the Adjudication of Federal Reserved Water Rights." Utah Law Review 1977 (No. 2):315-29.

Gavande, S. A., W. F. Holland, and C. S. Collins. 1978. Survey of Technological and Environmental Aspects of Wet-Residue Disposal in Evaporative Holding Ponds, Final Report. Austin, Tex.: Radian Corporation.

General Electric Corporation, TEMPO. 1973. Polluted Groundwater: Some Causes, Effects, Controls, and Monitoring. Springfield, Va.: National Technical Information Service.

Geswein, Allen, J. 1975. Liners for Land Disposal Sites: An Assessment. Washington, D.C.: U.S., Environmental Protection Agency.

Gill, Douglas. 1977. "Man, Nature Share Blame for Colorado River's Salinity." Denver Post, April 24.

298

Gill, Douglas. 1979a. "Lamm Decision About Salinity 'Difficult' One." Denver Post, September 16, p. 1.

Gill, Douglas. 1979b. "Saline Control Idea Blow to Diverters?" Denver Post, September 16, p. 15.

Gillette, Wyoming, Municipal Government Staff. June 1979. Personal communication.

Gilliland, Martha W. December 1979. Personal communication.

Gilliland, Martha W. 1981. "The Benefits of Water Management in Energy and Agricultural Development in the San Juan River Basin of New Mexico." Forthcoming in Resources and Conservation.

Gilliland, Martha W., and Linda B. Fenner. 1979. Alternative Water Management Strategies for the San Juan River Basin of New Mexico. El Paso, Tex.: Energy Policy Studies, Inc.

Glenn, Bruce, and Kenneth O. Kaufman. 1977. "Institutional Constraints on Water Allocation." Paper presented at the Energy, Environment, and Wild Rivers in Water Resource Management Conference, Moscow, Idaho, July 6-8.

Gold, Harris, Water Purification Associates. January 1980. Personal communication.

Gold, Harris, et al. 1977. Water Requirements for Steam-Electric Power Generation and Synthetic Fuel Plants in the Western United States. Washington, D.C.: U.S., Environmental Protection Agency.

Gold, Harris, and Calvin Calmon. 1980. Agricultural Water Practices in the West. Cambridge, Mass: Water Purification Associates.

Gold, Harris, and D. J. Goldstein. 1979. Wet/Dry Cooling and Cooling Tower Blowdown Disposal in Synthetic Fuel and Steam-Electric Power Plants. Washington, D.C.: U.S., Environmental Protection Agency.

Gray, D. E., New Mexico, State Engineer. June 1979. Personal communication.

Grimshaw, Thomas W., et al. 1978. Surface-Water and Ground-Water Impacts of Selected Energy Development Operations in Eight Western States. Austin, Tex.: Radian Corporation.

Hanke, Steve H., and James Bradford Anwyll. 1980. "On the Discount Rate Controversy." Public Policy 28 (Spring 1980):171-83.

Harbert, H. A., III, and W. A. Berg. 1978. Vegetative Stabilization of Spent Oil Shales. Washington, D.C.: U.S., Environmental Protection Agency.

Hartman, L. M., and Don Seastone. 1970. Water Transfers: Economic Efficiency and Alternative Institutions. Baltimore: The Johns Hopkins Press for Resources for the Future.

Hibbert, Alden R. 1979. Vegetation Management for Water Yield Improvement in the Colorado River Basin. Fort Collins, Colo.: U.S., Department of Agriculture, Forest Service, Rocky Mountain Forest and Range Experiment Station.

High Country News 12 (March 7, 1980): 13.

Hinds, Eugene. 1980. "Progress Report: Appraisal Investigation, Saline Water Use and Disposal Opportunitites, Colorado River Water Quality Improvement Program." Boulder City, Nev.: U.S., Department of the Interior, Water and Power Resources Service, Lower Colorado Region.

Hounslow, Arthur, et al. 1978. Overburden Mineralogy as Related to Ground-Water Chemical Changes in Coal Strip Mining. Springfield, Va.: National Technical Information Service.

Howe, Charles W., Professor of Economics, University of Colorado at Boulder. April 21, 1977. Personal communication.

Hu, M. C., and G. A. Englesson. 1977. Wet/Dry Cooling System for Fossil-Fueled Power Plants: Water Conservation and Plume Abatement. Research Triangle Park, N.C.: U.S., Environmental Protection Agency. As cited in Gold, Harris, and D. J. Goldstein. 1979. Wet/Dry Cooling and Cooling Tower Blowdown Disposal in Synthetic Fuel and Steam-Electric Power Plants. Washington, D.C.: U.S., Environmental Protection Agency, p. 52.

Humm, William R., and Edward Selig. 1979. "Water
    Availability for Energy Industries in Water Scarce
    Areas: Case Studies and Analysis." Unpublished
    report prepared for U.S. Department of Energy.

Idaho, Department of Water Resources. 1980. "Water
    Supply Bank Rules and Regulations." Unpublished.

Idaho Code, 1979a, §42-1766.

Idaho Code, 1979b, §42-1761.

Interagency Task Force on Irrigation Efficiencies.
    1979. Irrigation Water Use and Management.
    Washington, D.C.: Government Printing Office.

Interagency Task Force on Irrigation Efficiencies,
    Technical Work Group. 1978. Irrigation Water Use
    and Management, Review Draft. Denver: U.S.,
    Department of Agriculture, Soil Conservation
    Service, Cooperative Irrigation Study.

International Boundary and Water Commission. 1973.
    "Permanent and Definitive Solution to the
    International Problem of the Salinity of the
    Colorado River," Minute No. 242. Department of
    State Bulletin 69 (September 24):pp 395-96.

Josephson, Julian. 1980. "Safeguards for Groundwater."
    Environmental Science and Technology 14
    (January):38-44.

Kauffman, Ken, Water and Power Resources Service,
    Denver, Colorado. January 1980. Personal
    communication.

Kauffman, Ken, Bureau of Reclamation, Denver, Coloroado.
    June 28, 1977. Personal communication.

Kauffman, Robert F., Gregory G. Eadie, and Charles R.
    Russell. 1977. "Effects of Uranium Mining and
    Milling on Ground Water in the Grants Mineral Belt,
    New Mexico." Ground Water 14 (September-October):
    296-308.

Keller, Jack, Argricultural Engineer, Utah State
    University, Logan, Utah, as cited in Utah State
    University, Utah Water Research Laboratory. 1975.
    Colorado River Regional Assessment Study, Part 2:
    Detailed Analyses: Narrative Description, Data,
    Methodology, and Documentation. Logan: Utah State
    U., Utah Water Research Lab., p. 247.

Khoshakhlagh, Rahman. 1977. Forecasting the Value of
    Water Rights--A Case Study of New Mexico.
    Albuquerque: University of New Mexico, Bureau of
    Business and Economic Research.

Kiechel, Walter, Jr. 1975. "Inventory and Quantifica-
    tion of Federal Water Rights--A Common Denominator
    of Proposals for Change." Natural Resources Lawyer
    8 (No. 2):255-61.

Kirschten, J. Dicken. 1978. "The Quiet Before the
    Shootout Over 'The Water Law of the West.'"
    National Journal 10 (January 28): 149-53.

Kirschten, J. Dicken. 1977a. "Turning Back the Tides
    of Long-Time Federal Water Policy." National
    Journal 9 (June 11):900-3.

Kirschten, J. Dicken. 1977b. "Plunging the Problems
    from the Sewage Treatment Grant System." National
    Journal 9 (February 5):196-202.

Klarich, Duane A., and Jim Thomas. 1977. The Effect of
    Altered Streamflow on the Water Quality of the
    Yellowstone River Basin, Montana, Technical Report
    No. 3, Yellowstone Impact Study. Helena: Montana,
    Department of Natural Resources and Conservation,
    Water Resources Division.

Kleppinger, W. Michael. 1977. "Determination of
    Federal Water Rights Pursuant to the McCarran
    Amendment: General Adjudications in Wyoming."
    Land and Water Law Review 12 (No. 2):457-84.

Kurtz, Howie. 1979. "N.M., Ariz. Waters Probed for
    Effect of Radioactive Spill." Denver Post,
    September 2, p. 61.

Lane, George. 1979. "Water Board Sues to Bar Control
    Plan." Denver Post, July 26.

Larsen, Leonard. 1980. "Water Project Funding in a Fog." Denver Post, April 2, p. 20.

Laws of Utah, 1976, Chapter 23.

Leo, P. P. and J. Rossoff. 1976. Control of Waste and Water Pollution from Power Plant Flue Gas Cleaning Systems, First Annual R&D Report, EPA-600/7-76-018. Research Triangle Park, N.C.: U.S., Environmental Protection Agency.

McCarran Amendment (1952), §208 of the Department of Justice Appropriation Act of 1953, Pub. L. 82-495, Title II, 66 Stat. 560, 43 U.S.C. §666 (1970).

Maloney, Frank E., and Richard C. Ausness. 1971. "A Modern Proposal for State Regulation of Consumptive Uses of Water." The Hastings Law Journal 22 (February):536.

Maloney, Frank E., Sheldon J. Plager, and Fletcher N. Baldwin, Jr. 1968. Water Law and Administration: The Florida Experience. Gainesville: University of Florida Press.

Missouri Basin Inter-Agency Committee. 1971. The Missouri River Basin Comprehensive Framework Study, 7 vols. Denver: U.S., Department of the Interior, Bureau of Land Management.

Missouri River Basin Commission. 1978. Upper Missouri River Basin Water Availability Assessment for Coal Technology Requirements. Omaha, Neb.: Missouri River Basin Commission.

Montana, Department of Mines, Billings. April 1979. Personal communication.

Montana, Department of Natural Resources and Conservation (DNRC), Water Resources Division. 1976. Which Way? The Future of Yellowstone Water, Draft. Helena, Mont.: Montana DNRC.

Montana Code Annotated, 1978a, §85-2-102(2).

Montana Code Annotated, 1978b, §85-2-402(3).

Muys, Jerome C. 1971. Interstate Water Compacts: The Interstate Compact and Federal-Interstate Compact. Arlington, Va.: U.S., Natonal Water Commission.

National Academy of Sciences (NAS)/National Research Council (NRC). 1973. Weather and Climate Modification: Problems and Progress. Washington, D.C.: NAS.

National Petroleum Council (NPC), Committee on U.S. Energy Outlook. 1972. U.S. Energy Outlook. Washington, D.C.: NPC.

National Science Foundation (NSF), National Science Board. 1978. National Science Board Public Participation: A Regional Forum, Background Paper. Washington, D.C.: NSF.

Nevens, T. D., et al. 1979. Predicted Costs of Environmental Controls for a Commercial Oil Shale Industry, Vol. I: An Engineering Analysis. Washington, D.C.: U.S., Department of Energy.

New Mexico, Environmental Improvement Agency, Staff. 1977. Personal communication.

New Mexico, Water Quality Control Commission. 1976. San Juan River Basin Plan. Santa Fe: Water Quality Control Commission.

Nielson v. Newmeyer, 123 Colo. 189, 192, 228 Pac. (2d) 456 (1951).

North Dakota Surface Mine Reclamation Act, North Dakota Century Code, Chapter 38-14.1 (1979).

Northern Great Plains Resources Program. 1974. Water Work Group Report. Billings, Mont.: U.S., Department of the Interior, Bureau of Reclamation.

NRDC, Inc., et al. v. Train, 545 F. 2d. 320 (1976).

Odasz, Frank, Energy Transportation Systems, Inc. 1979. Personal communication.

Pernula, Dale. 1977. City of Gillette/Campbell County: 1977 Citizen Policy Survey. Gillette, Wyo.: Gillette/Campbell County Department of Planning and Development.

The Plains Truth. 1979. "Meanwhile, Back at the Ranch...." 8 (February): 4-5.

Price, Don, and Ted Arnow. 1974. Summary Appraisals of the Nation's Ground-Water Resources--Upper Colorado Region, U.S. Geological Survey Professional Paper 813-C. Washington, D.C.: Government Printing Office.

Pring, George, Environmental Defense Fund (EDF). January 1980. Personal communication.

Pring, G. W., and Karen A. Tomb. 1979. "License to Waste: Legal Barriers to Conservation and Efficient Use of Water in the West." In Rocky Mountain Mineral Law Institute, Proceedings of the 25th Annual Institute. New York: Matthew Bender.

Radian Corporation. 1977. An Investigation of the Potential for Utilization of Saline Ground Water in Energy-Related Processes, Final Report. Austin, Tex.: Radian.

Radian Corporation. 1978a. The Assessment of Residuals Disposal for Steam-Electric Power Generation and Synthetic Fuel Plants in the Western United States. Austin, Tex.: Radian.

Radian Corporation. 1978b. Surface and Groundwater Impacts of Selected Energy Development Operations in Eight Western States, Final Report. Austin, Tex.: Radian.

Radosevich, George E. 1978. Western Water Laws and Irrigation Return Flow. Ada, Okla.: U.S., Environmental Protection Agency, Office of Research and Development, Robert S. Kerr Environmental Research Laboratory.

Reclamation Act of 1902, Pub. L. 57-161, 32 Stat. 388.

Reclamation Project Act of 1939, Pub. L. 76-260, 53 Stat. 1187.

Resource Conservation and Recovery Act (RCRA) of 1976, Pub. L. 94-580, 90 Stat. 2795.

Rowe, Jerry W., and David B. McWhorter. 1978. "Salt Loading in Disturbed Watershed--Field Study." Journal of the Environmental Engineering Division, Proceedings of the American Society of Civil Engineers 104 (No. EE2, April):323-38.

Safe Drinking Water Act of 1974, Pub. L. 93-523, 88 Stat. 1660.

Saile, Bob. "Minimum Stream Flows Sought." Denver Post, January 20, 1977.

Salt River Valley Water Users Association v. Kovacovich, 411 F. 2d 201 (1966), 3 Arizona app. 28.

Schmidt-Collerus, Josef J. 1974. The Disposal and Environmental Effects of Carbonaceous Solid Wastes from Commercial Oil Shale Operations. Denver, Colo.: University of Denver, Research Institute.

Science and Public Policy Program (S&PP), University of Oklahoma. 1975. Energy Alternatives: A Comparative Analysis. Washington, D.C.: Government Printing Office.

Science and Public Policy Program (S&PP), University of Oklahoma. 1981. Energy From the West: A Technology Assessment of Western Energy Resource Development. Norman: University of Oklahoma Press. (In press)

Shaw, Bill. 1976. Environmental Law. St. Paul, Minn.: West.

Sherman, Harris D. 1977. "The Role of the State in Water Planning, Research, and Administration." In Water Needs for the Future: Political, Economic, Legal, and Technological Issues in a National and International Framework, edited by Ved P. Nanda, pp. 225-29. Westview Special Studies in Natural Resources and Energy Management. Boulder, Colo.: Westview Press.

Shupp, Liane E. 1980. "The Ultimate Chess Game of the Eighties." Colorado/Business 7 (January): 53-57.

Silverman, B., Water and Power Resources Service. 1980. Personal communication.

Simms, Richard A. 1980. "National Water Policy in the Wake of United States v. New Mexico." Natural Resources Journal 20 (January):1-16.

Skogerboe, G. V., and W. R. Walker. 1972. Evaluation of Canal Lining for Salinity Control in Grand Valley. Washington, D.C.: U.S., Environmental Protection Agency.

Smith, E. S. 1973. "Tailings Disposal--Failures and Lessons." In Tailing Disposal Today, edited by C. L. Aplie and G. O. Argall. San Francisco: Miller Freeman.

South Dakota Codified Laws Annotated, 1979 Supplement, §46-5-20.1.

Stockton, Charles W., and Gordon C. Jacoby, Jr. 1976. Long-Term Surface Water Supply and Streamflow Trends in the Upper Colorado River Basin, Lake Powell Research Project Bulletin Number 18. Los Angeles, Calif.: University of California, Institute of Geophysics and Planetary Physics.

Stone & Webster Engineering Corporation. 1979a. Solution Mining of Uranium: Administrator's Guide. Denver, Colo.: U.S., Environmental Protection Agency.

Stone & Webster Engineering Corporation. 1979b. Water Treatment Demonstration Facility: Sundesert Nuclear Plant Units 1 and 2, San Diego Gas and Electric Company--Final Report. N.p.

Strain, Peggy. 1977. "Water, Land, Life--It's All One in Valley Pipeline Debate." Denver Post, November 13.

Surface Mining Control and Reclamation Act of 1977, Pub. L. 95-87, 91 Stat. 445.

Susquehanna River Basin Compact (1970), Pub. L. 91-575, 84 Stat. 1509.

Thorpe, Michael. 1977. "Control of Water Within Indian Country--the Federal View Point." In Indian Law Conference, Phoenix, April 21-22, 1977: Materials for Registrants, Federal Bar Association. Washington, D.C.: Federal Bar Association.

Tipton and Kalmbach, Inc. 1965. Water Supplies of the Colorado River. In U.S., Congress, House of Representatives, Committtee on Interior and Insular Affairs. Lower Colorado River Basin Project. Hearings before the Subcommittee on Irrigation and Reclamation, 89th Cong., 1st sess., pp. 467-99.

Toxic Substances Control Act of 1976, Pub. L. 94-469, 90 Stat. 2003.

Trelease, Frank J. 1979. "Back to Basics--Taking the Politics Out of Water Law." Paper presented at Conference on Water Perspectives in the "Old West States," December 13-14.

Treaty between the United States of America and Mexico Respecting Utilization of Waters of the Colorado and Tijuana Rivers and of the Rio Grande, February 3, 1944, 59 Stat. 1219 (1945), Treaty Series No. 994.

Tweedell, Robert. 1980. Denver Post, March 16, p. 39.

U.S., Army, Corps of Engineers, Phoenix, Arizona. June 1978. Personal communication.

U.S., Commission on Intergovernmental Relations. 1955. A Report to the President [Eisenhower] for Transmittal to the Congress. Washington, D.C.: Government Printing Office.

U.S., [Hoover] Commission on Organization of the Executive Branch of the Government. 1955. Final Report to the Congress. Washington, D.C.: Government Printing Office.

U.S., Congress, Office of Technology Assessment. 1980. An Assessment of Oil Shale Technologies. Washington D.C.: Government Printing Office.

U.S., Congress, Senate. 1961. Report of the Select [Kerr] Committee on National Water Resources, pursuant to S. Res. 48, 86th Cong., 1st sess., January 30.

U.S., Congress, Senate, Committee on Energy and Natural Resources. 1978. Water Availability for Energy Development in the West. Hearings before the Subcommittee on Energy Production and Supply, 95th Cong., 2d sess., March 14.

308

U.S., Congress, Senate, Committee on Interior and
    Insular Affairs. 1975. Missouri River Basin
    Industrial Water Marketing. Hearings before the
    Subcommittee on Energy Research and Water Resources,
    94th Cong., 1st sess.

U.S., Congress, Senate, Committee on Judiciary. 1961.
    Delaware River Basin Compact, Senate Report 854, to
    accompany H. J. Res. 225, 87th Cong., 1st sess.,
    August 31.

U.S., Council on Environmental Quality (CEQ). 1975.
    Environmental Quality, Sixth Annual Report.
    Washington, D.C.: Government Printing Office.

U.S., Department of Agriculture (USDA), Soil
    Conservation Service (SCS), Special Projects
    Division. 1976. Crop Consumptive Irrigation
    Requirements and Irrigation Efficiency Coefficients
    for the United States. Washington, D.C.: USDA.

U.S., Department of Commerce, Bureau of the Census.
    1974. 1974 Census of Agriculture. Washington,
    D.C.: Government Printing Office.

U.S., Department of the Interior (DOI). 1973. Final
    Environmental Statement for the Prototype Oil Shale
    Leasing Program, 2 vols. Washington, D.C.:
    Government Printing Office.

U.S., Department of the Interior (DOI), Bureau of Land
    Management (BLM). 1976. Northwest Colorado Coal:
    Final Environmental Statement, 4 vols. Washington,
    D.C.: Government Printing Office.

U.S., Department of the Interior (DOI), Bureau of Land
    Management (BLM). 1977. Final Environmental
    Impact Statement: Proposed Development of Oil
    Shale Resources by the Colony Development Operation
    in Colorado, 2 vols. Denver: BLM.

U.S., Department of the Interior (DOI), Bureau of
    Reclamation (BuRec). 1972. Appraisal Report on
    Montana-Wyoming Aqueduct. Billings, Mont.: BuRec.

U.S., Department of the Interior (DOI), Bureau of
   Reclamation (BuRec). 1975. Westwide Study Report
   on Critical Water Problems Facing the Eleven
   Western States. Washington, D.C.: Government
   Printing Office.

U.S., Department of the Interior (DOI), Bureau of
   Reclamation (BuRec). 1977. Final Environmental
   Statement: El Paso Coal Gasification Project,
   San Juan County, New Mexico, 2 vols. Washington,
   D.C.: BuRec.

U.S., Department of the Interior (DOI), Bureau of
   Reclamation (BuRec). 1978a. Operation of the
   Colorado River Basin, 1977, and Projected
   Operations, 1978. Washington, D.C.: Government
   Printing Office.

U.S., Department of the Interior (DOI), Bureau of
   Reclamation (BuRec). 1978b. Status Report: Yuma
   Desalination Test Facility. Boulder, Colo.: BuRec.

U.S., Department of the Interior (DOI), Bureau of
   Reclamation (BuRec). 1978c. Water and Land
   Resources Accomplishments, 1977, Summary Report.
   Washington, D.C.: Government Printing Office.

U.S., Department of the Interior (DOI), Bureau of
   Reclamation (BuRec). 1979. Quality of Water,
   Colorado River Basin, Progress Report No. 9.
   Washington, D.C.: BuRec.

U.S., Department of the Interior (DOI), Bureau of
   Reclamation (BuRec), Upper Missouri Region. 1976.
   Water for Energy, Missouri River Reservoir, Draft
   Environmental Impact Statement. Billings, Mont.:
   BuRec.

U.S., Department of the Interior (DOI), Bureau of
   Reclamation (BuRec), Upper Missouri Region. 1977.
   ANG Coal Gasification Company: North Dakota
   Project: Draft Environmental Statement. Billings,
   Mont.: BuRec.

U.S., Department of the Interior (DOI), Bureau of
   Reclamation (BuRec) and Bureau of Indian Affairs
   (BIA). 1978. Report on the Water Conservation
   Opportunities Study. Washington, D.C.: U.S., DOI.

U.S., Department of the Interior (DOI), Bureau of
    Reclamation (BuRec); and Department of Agriculture
    (USDA), Soil Conservation Service (SCS). 1974.
    Colorado River Water Quality Improvement Program:
    Status Report. Washington D.C.: BuRec.

U.S., Department of the Interior (DOI), Bureau of
    Reclamation (BuRec), and Department of Agriculture
    (USDA), Soil Conservation Service (SCS). 1977.
    Final Environmental Impact Statement: Colorado
    River Water Quality Improvement Program. Denver:
    BuRec, Engineering and Research Center.

U.S., Department of the Interior (DOI), Office of the
    Solicitor. June 25, 1979. Federal Water Rights of
    the National Park Service, Fish and Wildlife
    Service, Bureau of Reclamation and the Bureau of
    Land Management, Solicitor's Opinion #M-36914.
    Washington, D.C.: DOI.

U.S., Department of the Interior (DOI), Office of Water
    Research and Technology (OWRT). 1980. Research
    Reports Supported by Office of Water Research and
    Technology Received During the Period October
    1979-March 1980. Washington, D.C.: U.S., DOI,
    OWRT, Water Resources Scientific Information Center.

U.S., Department of the Interior (DOI), Water and Power
    Resources Service (WPRS), Staff. 1980. Personal
    communication.

U.S., Department of the Interior (DOI), Water for Energy
    Management Team. 1974. Report on Water for Energy
    in the Upper Colorado River Basin. Denver: DOI.

U.S., Department of the Interior (DOI), Water for Energy
    Management Team. 1975. Report on Water for Energy
    in the Northern Great Plains Area with Emphasis on
    the Yellowstone River Basin. Denver: DOI.

U.S., Environmental Protection Agency (EPA). N.d.
    "Water Quality Impacts of Uranium Mining and Milling
    Activities in the Grant Mineral Belt, New Mexico."
    (Available through the New Mexico Environmental
    Improvement Agency, Santa Fe, New Mexico.)

U.S., Environmental Protection Agency (EPA). 1974.
    Clean Water. Report to Congress--1974. Washington,
    D.C.: EPA.

U.S., Environmental Protection Agency (EPA). 1979.
Water Quality Management, Five Year Strategey,
FY80-Baseline. Washington, D.C.: EPA.

U.S., Environmental Protection Agency (EPA). 1980.
"Water Quality Management, Five Year Strategy,
FY81-Baseline," Information Memorandum INFO 80-45.
Washington, D.C.: EPA.

U.S., Environmental Protection Agency (EPA), Office of
Drinking Water. 1979. Protection of Underground
Sources of Drinking Water: EPA's Underground
Injection Control (UIC) Regulations, Fact Sheet.
Washington, D.C.: EPA, Office of Drinking Water.

U.S., Environmental Protection Agency (EPA), Office of
Drinking Water, State Programs Division, Ground
Water Protection Branch. 1978. Surface
Impoundments and Their Effects on Ground-Water
Quality in the United States--A Preliminary Survey.
Washington, D.C.: EPA.

U.S., Environmental Protection Agency, Region VIII.
1980. "Energy Policy Statement." Unpublished
paper, Denver, Colorado, April 30.

U.S., General Accounting Office (GAO). 1979. Federal
Response to the 1976-77 Drought; What Should Be
Done Next?, Report CED-79-26 to the Comptroller
General, Washington, D.C.: GAO.

U.S., General Accounting Office (GAO), Comptroller
General. 1979. Colorado River Basin Water
Problems: How to Reduce Their Impact, Report
CED-79-11 to the Congress of the United States.
Washington, D.C.: Government Printing Office.

U.S., General Accounting Office (GAO), Comptroller
General. 1980. Water Supply Should Not Be an
Obstacle to Meeting Energy Development Goals, Report
to the Congress of the United States. Washington,
D.C.: GAO.

U.S., National Water Commission (NWC). 1973. Water
Policies for the Future, Final Report. Washington,
D.C.: Government Printing Office.

U.S., Office of Management and Budget (OMB), Study
    Committee on Policy Management Assistance. 1975.
    Strengthening Public Management in the
    Intergovernmental System. Washington, D.C.:
    Government Printing Office.

U.S., President's Water Policy Implementation. 1979.
    "Report of the Federal Task Force on Non-Indian
    Reserved Rights: Task Force 5-A." Unpublished
    report.

U.S., President's [Truman] Water Resources Policy
    Commission. 1950. A Water Policy for the American
    People, 3 vols. Washington, D.C.: Government
    Printing Office.

U.S., Water Resources Council. 1978. The Nation's
    Water Resources 1975-2000, Vol. 1: Summary, Second
    National Water Assessment. Washington, D.C.:
    Government Printing Office.

U.S. Code, Title 43, §372 (1976).

U.S. v. New Mexico, 438 U.S. 696 (1978), 98 S. Ct. 3012.

U.S. v. Tongue River Water Users Association et al.,
    Civil No. CV75-20-B19, August 1, 1975, U.S. District
    Court--Billings, Montana, Division, pending.

Upper Colorado River Basin Compact of 1948, Pub. L.
    81-37, 63 Stat. 31 (1949).

Uranium Mill Tailings Radiation Control Act of 1978,
    Pub. L. 95-604, 92 Stat. 3021.

Utah Code Annotated §73-3-8.

Utah State University, Utah Water Research Laboratory.
    1975. Colorado River Regional Assessment Study,
    4 parts. Logan: Utah State U., Utah Water
    Research Lab.

Valantine, Vernon E. November 5, 1980. Personal com-
    munication.

Van der Leeden, Frits. 1973. Groundwater Pollution
    Features of Federal and State Statutes and
    Regulation. Springfield, Va.: National Technical
    Information Service.

Van der Leeden, Frits, Lawrence A. Cerrillo, and David
    W. Miller. 1975. Ground-Water Pollution Problems
    in the Northwestern United States. Corvallis,
    Oregon: U.S., Environmental Protection Agency,
    Office of Research and Development, National
    Environmental Research Center.

Wade, James C., et al. 1977. "Sprinkle Irrigation
    Technologies and Energy Costs: A Comparative
    Analysis of Southern Arizona Irrigated Agriculture."
    Paper presented at the American Water Resources
    Association Conference, Tucson, Arizona, October 31-
    November 3.

Water Purification Associates (WPA). 1978. Aspects of
    Water Impact Analysis in Coal Conversion.
    Cambridge, Mass.: WPA.

Water Resources Planning Act of 1965, Pub. L. 89-80, 79
    Stat. 244, Title II.

Weisbecker, Leo W. 1974. The Impacts of Snow
    Enhancement: Technology Assessment of Winter
    Orographic Snowpack Augmentation in the Upper
    Colorado River Basin. Norman: University of
    Oklahoma Press.

Western Gasification Company (WESCO). 1974. Coal
    Gasification: A Technical Description. Farmington,
    N. Mex.: WESCO.

 estern Governors' Policy Office (WESTPO). 1978.
    Balanced Growth and Economic Development: A
    Western White Paper. Denver, Colo.: WESTPO.

Western States Water Council. 1979. Proceedings of the
    56th Quarterly Meeting, Keystone, Colorado,
    October 25.

Westinghouse Electric Corporation, Environmental Systems
    Department. 1973. Colstrip Generation and
    Transmission Project: Applicant's Environmental
    Analysis. N.p.: Westinghouse.

White, Irvin L., et al. 1979a. Energy From the West:
    Energy Resource Development Systems Report, 6 vols.
    Washington, D.C.: U.S., Environmental Protection
    Agency.

314

White, Irvin L., et al. 1979b. Energy From the West: Impact Analysis Report, 2 vols. Washington, D.C.: U.S., Environmental Protection Agency.

White, Irvin L., et al. 1979c. Energy From the West: Policy Analysis Report. Washington, D.C.: U.S., Environmental Protection Agency.

White, Michael D. 1975. "Problems Under State Water Laws: Changes in Existing Water Rights." Natural Resources Lawyer 8 (No. 2): 359-76.

Wild and Scenic Rivers Act of 1968, Pub. L. 90-542, 82 Stat. 906.

Willardson, Tony, Western States Water Council. November 1979. Personal communication.

Wilson, Leonard V. 1978. State Water Policy Issues. Lexington, Ky.: Council of State Governments.

Winters v. U.S., 207 U.S. 564 (1908).

Wyoming Statutes Annotated §41-3-105.

Wyoming Statutes Annotated, §41-3-115. (1977).

Yale Law Journal. 1979. "Indian Reserved Water Rights: The Winters of our Discontent." 88 (July):1689-712.

Yellowstone River Compact of 1950, Pub. L. 82-231, 65 Stat. 663 (1951).

Index